TECHNOLOGIES OF THE NEW REAL

Viral Contagion and Death of the Social

With astonishing speed, we have been projected into a new reality where interactions with drones, robotic bodies, and high-level surveillance are increasingly mainstream. In this age of groundbreaking developments in robotic technologies, synthetic biology is merging with artificial intelligence, forming a newly blended reality of machines, bodies, and affect.

Technologies of the New Real draws from critical intersections of technology and society – including drones, surveillance, DIY bodies, and innovations in robotic technology – to explore what these advances can tell us about our present reality, or what authors Arthur and Marilouise Kroker deem the "new real" of digital culture in the twenty-first century.

Technologies of the New Real explores the many technologies of our present reality as they infiltrate the social, political, and economic static of our everyday lives, seemingly eroding traditionally conceived boundaries between humans and machines, and rendering fully ambivalent borders between the human mind and simulated data.

ARTHUR KROKER is an emeritus professor and adjunct professor of political science at the University of Victoria. He is the director of the Pacific Centre for Technology and Culture (PACTAC).

MARILOUISE KROKER was a feminist scholar, publisher, editor, writer, theorist, and performance artist.

DIGITAL FUTURES

Series Editor: ARTHUR KROKER, University of Victoria

Digital Futures is a series of critical examinations of technological development and the transformation of contemporary society by technology. The concerns of the series are framed by the broader traditions of literature, humanities, politics, and the arts. Focusing on the ethical, political, and cultural implications of emergent technologies, the series looks at the future of technology through the "digital eye" of the writer, new media artist, political theorist, social thinker, cultural historian, and humanities scholar. The series invites contributions to understanding the political and cultural context of contemporary technology and encourages ongoing creative conversations on the destiny of the wired world in all of its utopian promise and real perils.

Editorial Advisory Board

For a list of books published in this series, see page 190.

Technologies of the New Real

Viral Contagion and Death of the Social

ARTHUR KROKER AND MARILOUISE KROKER

UNIVERSITY OF TORONTO PRESS
Toronto Buffalo London

ISBN 978-1-4875-4021-0 (cloth)   ISBN 978-1-4875-4024-1 (EPUB)
ISBN 978-1-4875-4022-7 (paper)   ISBN 978-1-4875-4023-4 (PDF)

Digital Futures

Library and Archives Canada Cataloguing in Publication

Title: Technologies of the new real : viral contagion and death of the social /
  Arthur Kroker and Marilouise Kroker.
Names: Kroker, Arthur, 1945–, author. | Kroker, Marilouise, author.
Series: Digital futures.
Description: Series statement: Digital futures | Includes bibliographical
  references and index.
Identifiers: Canadiana (print) 20210278757 | Canadiana (ebook) 2021027879X |
  ISBN 9781487540210 (cloth) | ISBN 9781487540227 (paper) |
  ISBN 9781487540241 (EPUB) | ISBN 9781487540234 (PDF)
Subjects: LCSH: Technology – Social aspects. | LCSH: Technology – Political
  aspects. | LCSH: Technology – Economic aspects.
Classification: LCC T14.5 .K76 2022 | DDC 303.48/3 – dc23

This book has been published with the help of a grant from the Federation
for the Humanities and Social Sciences, through the Awards to Scholarly
Publications Program, using funds provided by the Social Sciences and
Humanities Research Council of Canada.

University of Toronto Press acknowledges the financial assistance to its
publishing program of the Canada Council for the Arts and the Ontario Arts
Council, an agency of the Government of Ontario.

Dedicated to the memory of Marilouise Kroker, 1943–2018, writer, editor, poet of the spoken word, artist of the unspoken image, feminist theorist of bodies disavowed, prohibited, excluded, intellectual visionary of the digital future with its codes of domination and uncoded territories of resistance, imagination, and solidarity; and to our granddaughters, Claire and Lilith, who represent so well the brilliant promise of the rising generation of the young

Contents

Acknowledgments

Technologies of the New Real was begun in creative collaboration with Marilouise Kroker, who not only had superb insights concerning the fatal trajectory traced by contemporary technological innovations, including DIY bodies, surveillance practice, drones hunting humans, and robots in the age of artificial intelligence but also expressed her insights in both the writing of the different chapters as well as by brilliant poetry, including "Lip-Synching the Future," "Terror from Above," and "Night Sky Epilogue." While Marilouise did not live to experience the contemporary pandemic fever that is viral contagion and the death of the social, the power of her critical intellectuality touches every page of this book. It turns out that the intense collaboration that we shared over almost five decades, in writing, editing, performance, and publishing, definitely did not end with death, but very much continues with a form of writing that remains true to a deeply shared intellectual sensibility. Life may find its limits in the finitude of the body, but consciousness has a way of beginning anew in the limitlessness of the collaborative imagination, just that point where love, care for the other, and critical engagement with the world always finds ways of expressing itself.

I am grateful to the following people who have given me permission to use their material in the writing of this book: Dr. Deena Weinstein for permission to publish lyrics to "God Won't Bless America" and other quotes from Michael A. Weinstein's contributions to the music and manifesto of Vortis, a Chicago punk rock band of which he was both lead singer and theorist of the Vorticist imaginary; Craig Fahner for permission to publish excerpts from his insightful contribution to the online discussion of dystopia in the superb listserv, *-empyre- soft-skinned space* (https://empyre.library.cornell.edu/); Robyn Smith, editor, *The Tyee* (thetyee.ca) and Mitchel Anderson, author, "Humans Need a Prime Directive, Fast," *thetyee.ca*, 7 December 2008.

I am deeply appreciative of the admirable skills that Stephen Jones, my editor at the University of Toronto Press, has brought to the publication of this book as well as to his editorship of the *Digital Futures* series, combining creative editorial talent and very real intellectual rigour. I am also appreciative of the helpful suggestions and critical insights provided by anonymous readers of the original manuscript. Janice Evans and Carolyn Zapf at the University of Toronto Press have skillfully seen the manuscript through to publication.

During the isolation of the pandemic, my thought has been wonderfully nourished by so many writers, poets, artists, and theorists who have formed a continuing and sustaining intellectual community. My particular gratitude goes to Lynn Baron, poet, painter, and photographer, who has an eloquent way of making words listen again to the sounds of earth, sky, fire, and water; to David Cook who brings theory and art together into creative breakthroughs to the future; to Nichola Feldman-Kiss, critically engaged artist and social theorist, who draws into sharp visibility the bodies of the prohibited, excluded, and silenced; to Deena Weinstein, who jump cuts so easily from the sounds of heavy metal to the iron in the soul that is life today; to Francine Prevost, who has made of the pandemic an opportunity to rewrite the poetry of the aesthetic imagination; to Hy Mariampolski and Sharon Wolf for their postcard histories of the future that is our past; and, most of all, to my family, Alexis, Dave, Claire, and Lilith, who express care, grace, love, and wonderful reflections in this year of living solitude.

TECHNOLOGIES OF THE NEW REAL

Viral Contagion and Death of the Social

Introduction: Viral Contagion and Death of the Social

Winter of Hope/Spring of Melancholy

The disruption of everyday life has created an enormous opening that is quickly being seized upon by monopolistic digital platforms. Consumers have been driven into the waiting arms of Amazon, who happily takes human contact out of the equation for all manner of exchanges. Google and Apple pounce on the opportunity to develop contact-tracing technology, their products edging closer to becoming mandatory, rather than merely ubiquitous. And, in the absence of opportunities for physical human connection, platforms happily intervene, while maintaining practices that compromise user privacy and capitalize on user attention. If platforms were already on the path towards total integration into everyday life, then this very well might be the moment in which they consolidate their power over the imagination.

– Craig Fahner, *-empyre- soft-skinned space*[1]

During the pandemic, I sheltered in place on Vancouver Island, literally an island of attentive solitude in the global stream of viral contagion, thinking of technologies of the new real and the suddenly proximate, instantly changed meaning of "I Stepped into the Future and It Wasn't There." Here, the pandemic was controlled by means of a resilient public health system and political leadership deferring to medical expertise, relying on a widely shared sense of civic responsibility and general care for the community in responding to the COVID-19 virus. Now more than ever in this time of viral delirium, Vancouver Island seemed like a rare, magical intersection of the four meridians of air, earth, fire, and water, an undeclared republic tilting towards social justice just off the western continental mass of North America – social solidarity in the face of viral contagion.

Ironically, the winter months preceding the pandemic were just the opposite of isolation and social distancing. As part of a collective political struggle during the winter, many of us in British Columbia (BC) and elsewhere were involved in an active alliance with youth and elders involved in Indigenous resurgence and environmental activists to protest the armed occupation of Indigenous territories by the Royal Canadian Mounted Police (RCMP) in support of aggressive pipeline expansion. Like an epochal rip in the fabric of normal time and space, the provincial Parliament Buildings in Victoria, BC, were surrounded by a large Indigenous youth encampment, vibrant with the lighting of sacred fires, drumming, inspiring speeches, and a field of red dresses symbolizing murdered and missing Indigenous women, all carried out with a spirit of love, not violence, and with very courageous, very determined resolve on the part of the Indigenous youth and elders. I may have been teaching a seminar on the politics of race and power by day, with that haunting trilogy of *Black Skin/White Masks*, *Red Skin/White Masks*, and *Brown Skin/White Masks*,[2] but by night many students were at the encampment in active solidarity with Indigenous youth, while others responded to frequent appeals during the night-time hours to come to the Legislature to help protect the Indigenous youth from possible police violence. What I witnessed over the winter was a glimpse into the possibility of a more just future traced out in all its social creativity, political courage, and profound ecological understanding by Indigenous thought and practice, and by strong alliances between Indigenous youth and many other young people conscious of the historical injustices of settler colonialism.

Then, the pandemic struck, with all its globalized panic fear and political cynicism. The darkness of the pandemic spring was just the opposite of the lightness of winter politics. Watching President Trump's daily televised orgies of unconstrained narcissism and spasms of self-pity interspersed with mean-spirited scapegoating and cynical lies, all applauded by an enormous popular following howling the deep rage of their discontent, I was reminded of Deleuze and Guattari's description of the continuing power of seductive appeals to the suicidal death drive of fascism.[3] Here the political virus of right-wing populism, fuelled by panic fear and intense anxiety over the loss of jobs in the very real-life context for many people in the contemporary economy of work or starve, seeks to attach itself to the host cell of the television audience, releasing its genetic instructions and then waiting as the host cell reproduces the virus, whether expressed in the form of angry white male hysteria, scapegoating of Asians, border violence against asylum seekers, or studied popular silence concerning the cynical hijacking of relief funds by large corporations in the United States and by carbon-heavy energy companies in Canada.

The immediate consequences of viral contagion are dire: the eclipse of the social and the death of politics. And there's something else as well – something now present as a faint intimation of things to come on the horizon of perception but then quickly inflating into a really existent reality. That reality is *bio-fascism*. The signs are everywhere. A friend from New York texts me to express her concern about how quickly people are eager to surrender civil liberties in the face of the pandemic. She points to the *Wall Street Journal* with its reports on the alliance between Apple and Google in perfecting contact tracing. It's definitely a useful medical tool at the present moment, but after the pandemic, contact tracing is potentially a vast extension of the power of corporate surveillance over individual privacy for purposes of targeted relational advertising and, for the national security state, an emblematic breakthrough in power over the bodies of its citizens. Moralized first in the name of public health but later likely to be made permanent in the name of national security and virtual capitalism, contact tracing could well turn out to be a leading talisman of bio-fascism, with the workplace future likely to become an experiment in bio-politics – segregation of the population, temperature taking, sudden quarantines, rule by emergency decree. It's all of these things, while the virtual capitalism of the ruling financial corporations views this moment as a convenient time to actualize what has already taken place – the shedding of unnecessary living labour once commerce has fully transitioned to remote labour in the age of the gig economy. Here, surging gun sales and panic hoarding are only symptomatic signs of the death of the social and the eclipse of politics, and all of this disintegration moves to the background music of the coming of age of Bob Dylan's dirge, "Murder Most Fowl."

Along with these developments is the assumption that one lasting consequence will be to suddenly accelerate technological tendencies that, until now, have remained masked in society. Certainly, the impact of social distancing and physical isolation has volatized the appeal, if not everyday necessity, of online services: for shopping, shipping, and surfing. It's as if the social economy has suddenly split in two, with face-to-face businesses shuttered, slowed down with requirements for social distancing, or literally bio-pits for possible viral infections. The sudden silence in the streets of everyday business – stores, restaurants, malls, offices – contrasts sharply with the economic clamour in all the Amazons, Apples, and Googles of the digital world. Here, what appears to be happening is a decisive movement of the commodity form – the essence of all capitalist transactions – from the face-to-face market of the offline economy to the streamed services of the virtual economy. Following the doubled ideology of facilitation and command, virtual capitalism can so quickly triumph because, at this moment of very real social crisis, it easily facilitates the necessities and consumer choices of daily life, whether for food, services, banking, medicine, or entertainment. The question remains: after the pandemic, will what was so quickly and massively adopted by the population as a way of facilitating survival in the midst of social distancing become a new order of economy, less a matter of choice than a practical imperative imposed by the ruling virtual corporations? Will offices repopulate, or will business increasingly move essential services online? Will manufacturing plants seek to restore previous levels of employment, or will factories take advantage of this break in normal reality to speed up the transition long underway towards the automated future of labour promised by developments in artificial intelligence and robotics? In other words, what will be the future of life after the pandemic when the full adoption of technologies of the new real suddenly introduces us to a future as socially unexpected and unpredicted as it was technologically inevitable?

It is the very same with contact tracing. In this case, an innovative technology instantly deployable on all the mobile phone networks of the digital world to quickly facilitate the identification of patterns of viral infection is a perfect model of the technological ideology of facilitation and command. While some members of the public may express eagerness to sacrifice a measure of their own civil liberties for the greater good of public protection from a menacing virus, it will only be a small step for governments in active alliance with virtual corporations to insist that this technological innovation be installed in the operating systems of mobile communications. Contact tracing has been introduced as a necessary measure for better public health in a time of viral contagion. Who would be so churlish as to argue against such a morally sanctioned as well as medically useful proposal? But, of course, the potential consequence is that the real operators of the operating systems of mobile communications will have at their instant disposal a new and powerful gateway to political and economic perception. Like a long-desired, but never before implemented, instrument of granular political and corporate surveillance, contact tracing literally delivers the most intimate of bodily histories and their accompanying social networks to tracking, archiving, and surveillance by the dominant digital platforms of contemporary life. Speedily and ubiquitously delivered by the expedient of containing viral contagion, contact tracing is the core alphabet of surveillance of the future. Once the viral pandemic has subsided, as it inevitably will, the political benefits and economic incentives for platform politics and virtual capitalism will undoubtedly be too tempting to allow this innovation to slip away quietly. Perhaps maintained under the pretext of an early warning system for containing future epidemics or as an efficient instrument for fighting all the wars against terrorism that will surely populate future politics, contact tracing provides power with direct, unmediated access to the history of the digital self and its dense networks of social relationships. With this tool, the era of social media transforms itself into a medium accelerating the overexposure of the data selves that we have all become. Now, as always, the questions remain: Once contact tracing shifts from a technology of facilitation to a command software, *what* purposes will be served by contact tracing? *Who* will be the ultimate beneficiaries of this new method of soft-skinned control? *When* will it be activated as the new political attractor of everyday life? And, given the alliance of platform society with abiding forms of racial, sexual, ethnic, and gender divisions, *where* will it be applied and with what differential intensity? We may be rushing into the stormy darkness and bright sunshine of the twenty-first century, but it is wise to ask if contact tracing is not, in the end, only the newest digital version of the Trojan horse from Homer's *Odyssey*, but this time with invisible software algorithms of digital mythology replacing the hidden warriors of ancient Greek mythology. This sense of foreboding about life after the pandemic has begun to circulate in the digital stream, evident in discerning reflections by the contemporary artist Craig Fahner in his lucid contribution to a critical series of reflections on the pandemic by a global community of new media artists taking intellectual shelter in the "soft-skinned space" of the empyre listserv. For this contemporary artist, what is really at stake today in this era of technologies of the new real is nothing less than the triumph of platform capitalism versus the power of imagination.

> It is becoming clear that, with our increased dependence, the alienating qualities of life-via-platforms are laid increasingly bare. Boycotts of Amazon are becoming more widespread, as the working conditions in fulfilment centres and the activities of unions become more visible. The social media dreck funnelled into our filter bubbles, after months of screen time, feels completely meaningless. Calls for legislation that provides for human needs and collective health, rather than facile technological solutionism, are becoming louder. Every day at 7:30PM, my neighbourhood erupts with the sound of clanging pots and pans – a ritual I'm sure many are familiar with, meant to give thanks to frontline workers during the pandemic. I've come to see this ritual as not only a sign of appreciation, but as a gesture of unmediated togetherness that finds its way into everyone's closed-off domestic world. It serves to poke holes in the highly individuated bubbles that platform capitalism thrives on. If there is a moment in which refusal of the alienating tendencies of platforms is desired – in which public imagination might be channelled towards more meaningful forms of co-presence, exchange and resource distribution – then perhaps this is it.[4]

"God Won't Bless America"

It is Independence Day in America, 4 July 2020, just three days after Canada Day in my political neck of the woods, and I am thinking about where it all went so wrong and sometimes so right for these missionary experiments in settler colonialism across the Indigenous lands of Turtle Island. From force of habit, I attune my perception by reading once again the Book of Revelation with its biblical prophecy of a violent coming Armageddon, once all the signs have been fulfilled, while keeping another eye on Hannah Arendt's *The Origins of Totalitarianism*, with its hauntingly perceptive analysis of fascism in the lonely, disconnected age of mass society.[5] Watching images of social wreckage float by on the media stream – today it is Florida retirees giving a shout-out to "White Power" from their golf carts and wealthy suburbanites in St. Louis, Missouri, pointing loaded guns at anti-racism protesters; tomorrow it will be other passing scenes of the gathering social ruins – I turn up the volume on my ear buds, which are tuned to the sounds of "God Won't Bless America." Performed at raucous levels by a punk rock band, Vortis, from the heart of the heartless country of the American Midwest, it is laying down some heavy tracks about the American phantasmagoria as the bill comes due for the pleasures of the feast:

> Tell me what is the answer to the troubles of our
> Times
> We're reaping what we sow, we're paying for our
> Crimes
> {Chorus}
> God Won't Bless America again
> We're on our own, we're all alone
> The voices on the sky, only they won't die
> We're on our own, we're all alone
> We live in the orgasmatron, the outstretched
> Scratching hand
> The power of the claws, the law of the land
> {Chorus}
> In the land of the sea, the home of the scared
> The cages will be shut, and no one will be spared
> We're on our own. We're all alone.[6]

Everywhere the mood in the streets is grim. Everyday beach parties get giddier, city crowds more crowded, faces unmasked, hostilities more open, aggression frequent, denial all the rage, as a kind of plague frenzy takes possession of society. I read the manifesto for Vortis as an attunement to the psyche of the times. Following in the tradition of those other lonely hearts of renegade philosophy and art – Artaud and Nietzsche – this punk rock band takes its name from Vorticism, a brief, but immensely consequential, rupture in art history (1914–15) just before everything was crushed by the long slaughterhouse of the First World War.[7] In that moment before technologies of mass violence became the signature song of the twentieth century, artists and writers, including Wyndham Lewis and Ezra Pound, saw the dynamic energy of the vortex in everything. As the manifesto for the band states, "Pound's master image of the vortex is an INVERTED ELECTRIFIED CONE with an axis through the center, spinning furiously as it emits brilliant sparks of light."[8] Whirlpools, tornados, turbines, cyclotrons for sure – but now why not see the vortex as the way in which energy fast circulates in every-day media? The manifesto makes the point: "A physical vortex organizes the energies of the world through a whirling circular motion, forming an axis at its center, a vacuum whose force of attraction sucks the world into it and then spews it out in a continuous maelstrom."[9] It's just like the contemporary storm of politics, society, and economy in the era of new communication technologies: unpredictable whirling motions of energy moving in fast circulation, sucking the world into its vacuum by its force of attraction and then spitting out the difference. And, there's not just one dominant vortex, but multiple vortices in the media stream moving simultaneously, sometimes intersecting one another, at other times like whirling motions of energy independent, estranged, just seeking to vac-uum up all the particle energies of the world – and us with it – as just so many shards of broken glass, body clouds of the social ruins within, as we are drawn into the axes of all the vortices that move around us and through us every moment of the streamed media day. No one has im-munity: certainly not the prevailing mood of mass society, as individual psyches are overstressed, overexposed, and overstimulated by the rise and fall of anti-racism protests and violent police blowback; definitely not asylum seekers, migrant farm workers, and anyone below the pov-erty line, as their lives are sucked into a storm not of their own mak-ing and certainly not under their control and then expelled as so much waste material; and most clearly not the new pilgrims of the digital way, who are attracted by the whirling energy of new media only to find their nervous systems being quickly amputated by the powerful surgical force of artificial intelligence, granular surveillance, automation of just about everything, and the mesmerizing vacuum force of social media.

The contemporary social scene, then, is under the sign of the ec-stasy of catastrophe: drift culture in the "inverted electrified cone" of technological society, "with an axis going through its center, spinning furiously as it emits brilliant sparks of light." Trapped in the high-energy waves of the future opening before us as a horizon of violent vortices, we really have some clear choices: passively drift like so much digital wreckage – data debris in the maelstrom – waiting to be sucked into the energy field of the technological storm all around us or actively adopt our own variations of the strategy of this punk rock band from Chicago, who view themselves as Vorticists, "assimilating as many energies of the world as possible, feeling it deeply and intensely, and expressing their personal response to its wild diversity"[10] with discipline, energy, and creativity. For me, at some deep and guiding interior moment, that sounds just about right. All around us today, we can just hear the crash of implosive energies released first by individuals, then whole societies, seemingly giving up on life and making their peace with the death instinct that animates the unfolding story of contemporary technology. Call it what you will – the outsourcing of work in the age of remote communication, the externalization of consciousness that is streamed media, empathy mapping for a society that has projected its emotional life into the silent rustle of algorithms, soft-skin bodies, and waiting for the robots. As a collective species, we seem to have abandoned the complicated wrinkles of life, with its complexities and entanglements, for the promised peaceable kingdom of the machines. Maybe it was the continuous history of massive warfare violence that has tracked us since the First World War, the globalizing slaughter of the Second World War, haunting us with its memories of concentration camps and mass death in the heart of twentieth-century Europe and flaring up with wars that never end, some brushfire warfare, others whirling contagions. Or perhaps the always present vulnerability of that modern social invention, the self not the soul, just literally cannot take the stress of radical overexposure in every surface and orifice to always aggressive technologies seeking to inhabit consciousness, affectivity, perception, and labour. Whatever the reason, to speak now of the death of the social and the eclipse of the individual in a coming age of neural modification, massive data expropriation, privacy abuse surveillance, and the deep learning algorithms of artificial intelligence is only to make a cold-eyed clinical report of the contemporary state of events. And, if you catch my drift, we just might have an important choice to make before finally disappearing into the surrounding sea of wired circuitry: the pleasant amnesia of drift or something like the artistic strategy of Vorticism – deep immersion in every passing whirling axis of energy, no fixed position in advance, only fast responses with the aim of releasing the pent-up energy within into "brilliant sparks of light," and, most definitely, being multiple, variable, probabilistic, contradictory, with minds, half-data/half-human, moving at particle speed. This last is my recipe, at least, for learning how best to respond at particle speed to lip-synching the future in the age of technologies of the new real.

Lip-Synching the Future
We're drowning in cheap data
With no right to forget
Brains rewired
Bodies recoded
Computer logic is our only logic
Analytics is how we (now)
See the world

Technologies of the New Real explores the human impact of technology in the twenty-first century. Here, four critical intersections of technology and society – drones, surveillance, do-it-yourself (DIY) bodies, and recent innovations in robotic technology – are explored for what they have to tell us about the "new real" of digital culture. With astonishing speed and relatively little public debate, we have suddenly been projected into a new reality of pervasive surveillance, drone warfare, DIY bodies as the essence of the "quantified self," and creative developments in robotic technologies that merge synthetic biology, artificial intelligence, and the design of articulated robotic limbs into a newly blended reality of machines, bodies, and affect. However, while the sheer dynamism of this digital remaking of human experience seemingly anticipates a future of accelerated technological change, it does not account for the dark singularities of increasingly atavistic politics, fatal flaws in the codes, the "blowback" of long-suppressed ethnic and racial grievances, or the rise of fundamentalist ideologies.

Technologies of the New Real seeks to answer the question posed by the uncertain world of twenty-first-century experience itself: namely, why in an age of a seemingly inexorable drive to technical perfection, smart bodies, and complex machine-human interface has society itself so quickly imploded into politics moving at the speed of darkness and motivated by the will to purity? Consequently, a truly unique world situation has come about, featuring powerful eruptions of the boom and bust cycles of late capitalism; the rise of reactionary fundamentalist movements, some religious, others political; the effective political dispossession and economic destitution of significant portions of the world's population; and yet, in the midst of all of this, the emergence of a new technological theology as transcendental in its cosmological ambitions as it is localized in its implications. So, then, we live in a twenty-first century that may have permanent war, class privilege, and resurgent forms of political recidi-vism as its sustaining noise. But, for all that, there is the clear signal in the technological background of ambient robots, DIY bodies, hovering drones, and machine-readable mass surveillance that something else is happening, something as novel in its technical expressions as it is enig-matic in its consequences. *Technologies of the New Real* is about listening intently to the signal of technologies of the new real as they penetrate the social, political, and economic static of the post-human condition, seemingly erasing traditionally conceived boundaries between humans and machines, and rendering fully ambivalent borders between minds and data flows.

In this case, every technological device is a symptomatic sign of the times with a complicated story to tell. Crystallizing the creative ener-gies, market-driven demands, popular desires, and acute anxieties of the society that surrounds them, technological devices – smart phones, iPads, tablets, 5G screens – implode in society like powerful singular-ities: creating new digital futures, quickly sidelining the past, silently reframing narratives of individual lives, dividing generations, relentlessly embedding cybernetic values, accelerating the rise of the privileged technocratic class, and equally, at key points, making unnecessary, if not impossible, the labour of those who do not enjoy the warm sunshine of technocracy with its futurist arcadia of deep machine learning, artificial intelligence, and labouring robots. For all of its truly wonderful magical qualities and spectacular powers of communication, no technological device is born innocent. In what is often considered to be something admirably non-political – a means not an end of communication, a facil-itator not a demander – technology in general and technological devices in particular are actually how we really experience what is publicly val-ued in contemporary society: living with data, connectivity, speed, being in the loop, blowing up and going viral, an influencer or a follower, a maker of memes, a gamer of all things digital.

Hyped by multinational network conglomerates as a sure and certain path to economic opportunity and creative freedom, feared by some for their tangible hints of a greater dependency, technologies of the new real are always complex in terms of their interests; complicated in their expression of values that they silently nest in consciousness, imagination, and feelings; and always enigmatic in their consequences. Not so much a matter of intended or unintended consequences, the story of technol-ogies of the new real – the digital horizon that is the air we breathe, the data we surf, the screens we inhabit – is itself intensely consequential. For better or worse, for purposes of personal survival, for the pleasures of entertainment, or for the daily routine of compulsory screen time at the office, often for relief from boredom, we as a society are deeply caught up in the greater adventure of technology. Today, technology deeply shapes the question of identity itself: who we are and what we would like to become, certainly as individuals but also as members of political entities and society as a whole. But, if that is the case, what is our technological identity? Is it lip-synching the future, drowning in cheap data with no right to forget, brains rewired, bodies recoded? Or is it something else – critically engaged individuals turning directly into the storm of technology in order to better understand its contradictory tendencies towards creative freedom and digital injustice? Of course, turning into the data storm – understanding the question of technologies of the new real – can only really begin with an understanding of the larger context out of which grow technologies of the new real: smart bodies, ubiquitous surveillance, hovering drones, and robots resurgent. If there is a palpable sense today that technologies of the new real are, in effect, transitions to an unfolding digital future, the dimensions of which are still clouded, still uncertain in detail and definition, perhaps that sense reflects a growing uncertainty and palpable anxiety about the times in which we live, a sense of stepping into a future that isn't there.

Scenes from the Future: Four New Trajectories of Social Media

Consider this insightful reflection on the era of technological disruption by the media commentator Mitchell Anderson:

> Climate is perhaps the most pressing crisis looming in the near future, but it is by no means the only one. A rogue Chinese scientist reportedly just genetically modified two babies using widely available and cheap CRISPR gene editing technology. The descendants of such newborns will pass on their engineered mutations to future generations and whatever inadvertent surprises nature might have in store. What could go wrong?
>
> Silicon Valley drives societal transformation through a suite of addictive and disruptive technologies already upending everything from courtship to cab driving. Some observers rightly worry the secretive and unregulated technological arms race towards artificial intelligence (AI) will one-day end very badly. In the meantime many of us are enjoying the pleasantly warm phase of the frog-boiling exercise even as the transition to a high-tech world exacts a mounting human toll.
>
> Up to 800 million jobs may be eliminated by 2030 to automation displacing occupations from the factory floor to brain surgery. Blue-collar workers are so far bearing the brunt of this transformation, driving droves of frightened and angry underclass into the arms of political populists. In

the longer term, a profit-driven dash towards ubiquitous AI means none of our jobs may be safe. Are we philosophically prepared for the future? Or the present?

Coping with normal is hard enough without asking people to adapt to an existence they can't even imagine. What would a world without fossil fuels even look like? People in such unsettling times are understandably clutching closer to culture and class as the ground beneath shifts. The *gilets jaunes* revolt in the streets of Paris is named for the yellow safety vests of the lower-wage workers most affected by rising fuel prices. Political leaders in Alberta are falling over themselves to seem the most slavishly devoted to the oil industry and present day workers rather than future generations.[11]

In his reflections on technologically induced change – climate change, genetic modification, AI – Mitchell Anderson poses an important question concerning the digital future: now that leading societies of the West have moved decisively from religion to capitalism as their central organizing principle, from concern over the fate of the soul to our identity as consumers, what happens when capitalism itself falters as an organizing principle under the pressure of the onrushing future? After souls and consumers, what will be the new "prime directive," the new ethical vision, which will guide us through the Scylla and Charybdis of this age of technological disruption?

This question is an important one precisely because it captures the essence of what is at stake in thinking through trajectories of the future. While the twentieth century might have begun with Nietzsche's fateful prophecy concerning the death of God and the rise of the human finally coming into its own subjectivity – whether benign, creative, or murderous – the twenty-first century surely begins with another epochal break, namely the death of the human and the rise of hybrid bodies in a fully blended reality of clones, drones, and androids, with time itself suddenly moving simultaneously in three wildly different directions. Everywhere now, we hear politically powerful and deeply reactionary appeals for an idealized, nostalgic (white) past that probably never existed. At the same time, in all the creative digital and biogenetic labs generating the technological future, messianic movements are on the move, mesmerized by dreams of technological rapture, that point where humans finally gain the long-sought goal of immortality by merging with machine culture – the so-called singularity moment. The present time of the contemporary human condition is scavenger time, where the present seems to live increasingly on the remains of the death of truth, the death of the social, the death of law – a time of kitsch and decay, growing cynicism, and sad remainders of the day with multiple signs of sign slides moving the human condition forward. More than is realized, we may be living in the "onrushing future" first envisioned by the French theorist Paul Virilio, who predicted that the "information bomb" has "accidented" not only traditional understandings of time and space in favour of the real time and virtual space of digital reality but that technology may have actually accidented the meaning of the human itself.[12] Of course, probably shocked by this impossible meeting of past, present, and future, the technologically accidented body has done exactly what was prophesied sixty years ago by Marshall McLuhan, a culturally attuned observer of the onrushing future. The body has shut itself down for survival purposes. And well it might, since the eclipse of the human, this signal event that will serve as the horizon of the twenty-first-century digital future, has been accelerated by four fundamental drives contained within the will to technology: the *externalization* of human consciousness, the *objectification* of the human subject, the *synchronization* of human emotions, and the *virtualization* of culture. Considered individually, each of the above technological drives has the effect of transitioning a specific dimension of human experience – mind, subjectivity, affect, and values – to full integration with the power of flows characteristic of network culture. Considered together, the combined power of these animating technological drives intimates a digital future where the eclipse of the human is paralleled by the new synergies of living with data. Here, ablated consciousness fit for a future of distributed knowledge, data flesh perfect for circulation as tracked objects in the Internet of Things, synchronized emotion for quicker mobilization in the key drives of the technological future, and the virtualization of the human cultural imagination itself are key trajectories in the digital future.

First Trajectory: Questions from the Shallows – Distributed Cognition

Named after the brain molecule that gives us pleasure, Dopamine Labs uses computer coding to influence behaviour – most importantly, to compel people to spend more time with an app and to keep coming back for more.

Co-founder Ramsay Brown, who studied neuroscience at the University of Southern California, says it's all built into the design, namely that, in the contemporary era, "we're not just designing software anymore, we're designing minds."[13]

The core trajectory of the digital future is the externalization of human consciousness, the movement of cognition from its biological habitat in an individuated brain to an era of distributed knowledge where everyone is free to share the digital commons – an increasingly common data mind moving at light speed. Here, the distributed world of conscious-ness, from politics and economy to culture, no longer remains at arm's length from the individuated brain but begins to engage in a complex feedback loop with consciousness itself, providing flows of real-time data, images, sound, and commentary that are intended to have the effect of erasing boundaries between mind and the outside world, ideally making of everyone's brain a fast flowing vector in the media stream. First pre-dicted by Marshall McLuhan with his conceptualization of "ablated" con-sciousness as a key trajectory of electronic media of communication,[14] the externalization of human consciousness has been carefully tracked by contemporary media theory, from Paul Virilio's concept of "duplex consciousness" and Jean Baudrillard's culture of "simulation" to Michael A. Weinstein's theorization of the externalization of the mind as one of the core features inaugurating the era of the eclipse of the human.[15]

Certainly not a bleak Orwellian 1984 scenario where individual con-sciousness is brought under the control of the surveillance state, the drive to externalized consciousness is transcendental in its aspirations, aimed, at least in its inaugurating rhetoric, at enhancing creative free-dom. Here, rather than the life of the mind continuing to be conceived as an innate biological property of individuals, the gateways of digital perception suddenly open to allow mind merger with the world of dis-tributed knowledge, surfing the data stream, connecting personal narra-tives with public autobiography, saturating the digital self fully in flows of interesting, sometimes necessary, information, stories, and feelings, but also actually putting flesh on the digital self by the reverse move-ment that takes place when distributed knowledge migrates from the outside of the brain to its cognitive interior, from something external to the human mind to something suddenly deeply resident in its neurolog-ical operation, from an occasional amusement to an addictive habit – the digital brain in the (wireless) stream: animated by data, powered up by the speed of perception, captivated by the rhythm of the flow, motivated by being the architect of its own digital destiny, seduced by the startling intimacy of a data world that makes up for a big fall in consciousness by an event of the bigger rise in connectivity.

In other words, the life of the individuated mind may already be peripheral to the story of minds as machines come alive. That's the story of artificial intelligence and deep learning, which, while they may have begun as poor proxies for human consciousness, have quickly seized command (cognitive) positions in the digital future as the privileged form of distributed knowledge. While we will increasingly recognize their deep integration in society – driverless cars, automated factories, virtual financial advisors, algorithmic networks moving at the speed of deep learning – what is less apparent, but more decisive, is that the appearance of artificial intelligence and deep learning are breakpoints in the history of human consciousness, signalling that moment when life in the wires means coming to terms with the automation of the mind. Not in a necessarily conformist way as in traditional scenarios of mind control, this coming to terms is a much more complicated merger of the true complexity that is human consciousness with the deep learning of virtual machines. It is one of those epochal events, then, when things are definitely in the balance, when machines, serving as feedback loops for human minds hungry for hits of information, suddenly have to deal with all the creativity, perversity, and beautiful strangeness and unpredictability of the fully blended mind rising for the first time into historical presence. Here, the eclipse of the human as symbolized by the ablated mind is, perhaps, just the beginning of the phantasmagoria of shared consciousness, the adventures in cognition that are algorithms made flesh, deep learning undone by very human questions from the shallows.

Yet, the question of politics is never far away from the universe of externalized consciousness. Indeed, it is decisive in an era of distributed cognition control, not only of the contents of distributed thinking but, more critically, control of the form of externalized consciousness – control, that is, of the complex network infrastructure of high-speed digital communication, which is the backbone of distributed consciousness. That is why the following story is of more than local significance, marking as it does one of the first cyber battles between China and the United States for control of the future of distributed consciousness.

THE NEW LONG MARCH OF 5G INFORMATION WARFARE

He who has the speed has the power.

– Paul Virilio, *Pure War*[16]

In 5G Race with China, US Pushes Allies to Fight Huawei

– *The New York Times*, 26 January 2019[17]

Alarmed by the resurgent global prominence of Huawei, a major privately owned Chinese telecommunication corporation now leading the world with its rapid deployment of a highly innovative, fast, and cheap 5G network information platform – the new network standard for speed, fluidity, and flexibility in moving vast streams of data – the suddenly challenged software empire of the United States has struck back hard and fast, deploying not innovative technological solutions but political stealth. At stake in this struggle between Huawei, as a surrogate for Chinese strategic national ambitions to quickly climb the economic value chain from low-wage mass manufacturing to broad spectrum technological innovation, and the United States government, as a surrogate for American network infrastructure companies that find themselves stuck in the mud of suddenly eclipsed 4G information systems, is the question, *Who will code the digital future?* Here, control of network technology involves powering up the information bomb that is the digital future, putting down codes for which economies, cultures, and politics matter and which will be silently but no less relentlessly prohibited, excluded, and silenced. In this scenario, California ideology – the technological vision of information technology Silicon Valley style, with its concerted orchestration of actions between Silicon Valley and the American national security state – finds itself challenged by a rising star on the previous digital periphery: Huawei, a company with a work force of engineers and researchers based in the Pearl River Delta north of Hong Kong. In digital reality, innovative code breakthroughs are always restless, know no definite boundaries, and move from the periphery to the centre at the speed of viral contagion; with this struggle, California ideology meets its first real challenge for control of the digital future in the form of the nervous breakthrough that is 5G at the coding hands of Pearl River ideology. It is an ideological struggle. Just as much as California ideology assumed a dominant global position in computer technology by first silencing that other California counter-ideology, those hackers of computers, politics, and art who dreamed of turning the magic of information technology in the direction of digital justice, so too Pearl River ideology runs parallel to the global ambitions of the Chinese state, namely to demand respect for China, for its past history as well as its present accomplishments, as an author of a future outside the orbit of American political dominance. So, then, it's a classic struggle between massive trading blocks, each of which shades into equally massive political empires. In this case, information *is* political, and there is nothing more politically sovereign than control over the nervous system of global information infrastructures.

China understands this reality, which is why it has staked its long-term political strategies on the new long march of codes. The United States does as well, which is why it has laid down three challenges to resurgent China. First, buried in the text of recent American tariffs against Chinese imports are specific regulatory clauses aimed directly at prohibiting the global adoption of Chinese computer infrastructure products. China might aspire to climb the economic value chain, from low-wage mass manufacturing for export to consumer markets to high-wage technological enterprises, but the United States is clearly determined to slow down, if not block entirely, this key code shift in Chinese state planning. Just as in the Cold War with the Soviet Union, where the United States swiftly put in place a policy of containment – the global American-led international alliance against the spread of communism – as a way of limiting the spread of socialism, so too in the now emerging cold (data) war with China, the first stage of digital containment involves constructing an alliance of willing nations – Japan, New Zealand, Australia, the United States, and potentially Canada and Britain – officially committed to *not* deploying Chinese computer infrastructure in government and military digital installations. Perhaps to make the message clear or perhaps just as a deliberate, secretive game between the digital masters of "Five Eyes" – the signal intelligence sharing agreements among five countries, Australia, New Zealand, the United States, Canada, and Britain – and enthusiasts for free trade with China, the arrest of the Huawei's chief financial officer (CFO) for legal extradition to justice American-style marks the beginning of overt hostilities between these two clashing digital powers, one resurgent, the other seemingly in danger of being technologically eclipsed.

In digital capitalism, network speed is everything. When information moves at light speed with stock exchanges, media outlets, military strategists, banking networks, and gambling sites responding real second by real second to dips and flows of information and with everything to be gained or lost by the ability of networks to do something about it, then the speed, fluidity, and flexibility of network technology is the essence of digital reality and most certainly the nerve centre of the digital commodity form. Now, as always, *new relations* of network communication necessarily render obsolescent *old forms* of network technology. In the battleground of 5G technology, 5G is the spearhead of the digital future, the new relation of network communication assured of global sovereignty by the logic of capitalism itself. Indeed, with strategies of technological innovation now more decisive than tactics of military power in the games of empire, information warfare fought code by code in the dark streets of network infrastructures is how the digital future will be won or lost. Consequently, the following early warning signals were issued from the front page of the *New York Times* Sunday edition, where it was argued that the United States is engaged in a renewed arms race "involving technology rather than traditional weaponry" and that what is at stake is the development of immensely powerful cyber weapons such as 5G, which will, in effect, shape the future of the twenty-first century. Here, the future is envisioned as powered by artificial intelligence (AI) and deep machine learning with networks specifically designed around the imperatives of "technological stacking" and ubiquitous computing, creating continuous data flows among cities, manufacturing production, and a supporting background structure of autonomous cars and trucks,

complex sensors, and self-directed robots.[18] Not just a future of autonomous weaponry but also increasingly autonomous (technological) life is
envisioned, with all of it moving at the speed of 5G.

Second Trajectory: Globally Synchronized Emotion

The French media theorist Paul Virilio once noted that a major political challenge for the twenty-first century has less to do with a future of
control over public opinion than with the synchronization of emotion
on a planetary scale. In this case, what is at stake is not so much the politics of issue management by the manipulation of public opinion but the
massive effect of social media in making possible the synchronization of
human emotion, often on a world scale. Here, it is not a matter of constructing mass clusters of public opinion but of the movement of human
affect into the deepest, yet most openly spectacular, dimensions of social
media. Of course, while Virilio worried about the possibility of mass synchronization of human emotions by controlling elites, what is more the
case today is the amplification on a global scale of emotions that, while
they may be synchronized in form, are radically dis-synchronized in content. The real effect of technologies of the new real has been both to
throw a bright spotlight on lingering human grievances – racial, ethnic,
class, sexual, nationalist – and to accelerate long-buried grievances with
the psychic charge of shared emotions, shared information, in social media as well as on mass media platforms. When previously private affect
discharges directly into the public scene, the political consequences are
as unpredictable as they are sometimes undesirable.

Consider the present political situation in which resurgent white nationalism is the storm centre of global politics in the twenty-first century,
from the political epicentre of the previously technological liberal American government to the rise of Eurasian ideology in Russia and right-wing,
anti-immigrant, white nationalist movements and governing powers in
Hungary, Italy, Spain, Poland, Canada, Britain, and France. Affectively,
the political formula of white nationalism is always the same: bitter nostalgia for the loss of white sovereignty in politics, economy, and culture,
with its psycho-affectivity in its deepest origins made all the more embit-
tered, angry, and anxious by the purely phantasmatical quality of its perceived threats – fantasies of terroristic border incursions, social collapse,
cultural contamination, economic crisis, impurities at the borders of
blood and territory – as psychological staging grounds for the exclusion
of refugees, asylum seekers, and immigrants. When the threatened society of the angry mass that is white nationalism seeks to take shelter from
the gathering storm, it turns inward only to discover an existential crisis

of an often lonely, disconnected, powerless, melancholic self. As Franz Neumann, a member of the Frankfurt School who had fled German fascism, noted, when the external crisis meets the internal crisis of the melancholic self, the result is fascism, with its search for a strong, author-itarian leader who will provide direction in the search for scapegoats to exact revenge for often-imaginary wounds.[19] Consequently, an ontol-ogy of hatred of existence becomes the essence of white nationalism for individuals who, as Nietzsche predicted long ago,[20] have grown sick and weary of themselves. Volatilized by social media and quickly circulating as globally synchronized emotion, the present fallout from the spread of the ontology of hatred of existence has an immediate political conse-quence: the rise to political power of ethically fatigued but charismatic authoritarian leaders whose political gift lies in creating a spiderweb of appealing narratives to bind together in the melancholic language of anger, rage, and revenge the global flow of white nationalism.

Like all previous primal hordes, white nationalism involves intense emotional involvement, serial thinking with a libidinal register, which takes its cue from strong leaders' aggregating melancholic egos into powerful political movements, providing sacrificial targets for spasms of ressentiment and engaging in a politics of hand-to-mouth daily strug-gle against bodies negatively designated under the signs of disavowal, prohibition, and exclusion. Here, the ontology of hatred of existence, once circulated in the form of synchronized emotional narratives by social media, will be all the more aggressive considering the practical and, indeed, psychological impossibility of constructing strong walls around the always chimerical illusion of pure white identity.

Third Trajectory: The Ecstasy of Objects – Part Data/Part Flesh

Digital devices always seem to follow the same logic, moving from the out-side of the body to its interiority, already operating as feedback loops in cognition, affect, and perception, becoming smaller, moving from visibil-ity to invisibility, attaching themselves to bodies in the form of seductive prosthetics. Everything and everyone it seems is already psychologically prepared for the next stage of technological evolution, that point where technologies of communication throw off even the masquerade of pros-thetic devices, becoming part of bodily biology itself. How it will happen is still to be decided, whether by chip implants at birth or perhaps by developments in biogenetics that allude to the possibility of telepathy.

However, what happens in a real world of DIY bodies *when we become the digital devices that we thought we were only using?* Relationships move to the rhythm of Snapchat, Grindr, and Tumbler, to Instagram for intimacy, while digital selves orbit around their own identity in the form of Facebook, with Twitter chats, emoticon emotions, and memories safely Google cached. At what point does the seduction of technological devices, with their powers of facilitation, projection, and augmentation, substitute the remainders of human subjectivity with dreams realized of becoming digital, of actually being the device of their dreams: thinking like an algorithm, seeing like a drone, monitoring like a Fitbit, imaging the world like a Netflix, our social identity assembled and reassembled like a technological autobiography, feeling like a digitally recoded, accelerated, powered up version of our previous offline self?

Probably more than we suspect, we have perhaps always been transhuman. Not transhuman in the current sense of anxiously, or perhaps eagerly, waiting for the "singularity moment" of technological dreams, but transhuman in the real-time sense of being fully transitional beings of flesh and data experiencing all the complexity and unpredictability associated with deep immersion in a universe of interconnected devices. Here, life in the wires means life in the Internet of Things, where previously silent objects – refrigerators, rooms, cars, trucks, computers – begin to speak, communicating with other devices in digital tongues, sometimes slurring their words when the network drops but, for all that, working hard to stabilize bonds of connectivity, streams of communication among human interlocutors and an interconnected world of smart phones, cars, airplanes, and household appliances. In the real-time world of connected devices, what matters is actually becoming a functioning relay in a complicated network of devices with deep learning as its pedagogy and artificial intelligence for a brain. Here, human subjectivity is at an evolutionary disadvantage if it cannot keep up to the speed, fluidity, and flexibility of digital communication. While in the beginning, it was probably sufficient to externalize the human nervous system by sacrificing human senses to their counterparts in the form of electronic media of mass communication (radio for an amplified ear, television for a prosthetic eye), today something else is required for the sacrifice, namely the disappearance of human subjectivity into an object of its digital desires. Like a satellite in orbit, subjectivity does not simply vanish from the scene but continues its rotation from a distance around its evolutionary successor species – data flesh.

Unlike all the grim prognoses of life as a digital object, the downloading of subjectivity into the digital devices that we have become is not apocalyptic. Quite the opposite. While the overall technological drive is towards the *full objectification* of everything – devices, humans, nature – the entry of the human into the world of objects is only the beginning of a new adventure. When the all-too-human will to power meets the will to technology, when subjectivity dares to walk across the digital abyss, the result is unexpected, namely the animation of the previously hygienic world of pure objects with all the necessary messiness of the human condition. When human noise meets digital signal, actually merging noise and signal in a complex knot of its own creation, multiplicity happens. Ironically, life as a communicative object in a real-time world fully objectified will probably result in undermining the artifice of the supposedly seamless world of objectification. After all, no one has ever really been interested in being thing-like, least of all objects themselves. Literally left to its own devices, the universe of communicative objects is always in danger of dying from a lack of energy, from an absence of spontaneity, unpredictability, perversity, complexity, from having to function so exhaustively to maintain the illusion of transparent, real-time communication. With its strong animating desires, its multiple tastes, moods, and anxieties, with its doubled appetite for nausea and rapture, the migration of human subjectivity beyond its own skin barrier by being digital will come to haunt the story of technology itself, making the real singularity what happens after human subjectivity transforms into its twenty-first-century counterpart: beings part data/part flesh. Not so much, then, will it be a once and future world of pure speed, but something else – a future being part data/part flesh accelerating to inertia. In this case, contrary to technical dreams of a fully objectified, fully connected digital universe, technologies of the new real will undoubtedly deliver something radically different, namely the growing importance of understanding technology as a liminal zone, an indeterminate space of the in-between, where the digital drive to full objectification meets the equally powerful human drive to integral subjectivity.

Fourth Trajectory: Accelerating to Inertia

All medical images contain some visual noise. The presence of noise gives an image a mottled, grainy textured or snowy appearance.
– Perry Sprawls, "Image Noise"[21]

Externalizing human consciousness, mass synchronization of emotions, subjects rising into objectivity: what are these but symptomatic signs of a fantastic acceleration of contemporary events moving to inertia? It's not inertia understood as a necessarily negative state of entropy but something else: inertia as the acceleration of historical events with such velocity and magnitude of data acquisition that the event scene which is contemporary society follows a violent trajectory that suddenly turns back on itself, abruptly reversing direction until finally tumbling into a floating, free-fall state of cultural weightlessness – a delirious, phantas-magorical crash zone where imaged minds, massaged emotions, and the body as prosthetic begin to float like clouds of data dust in the gathering storm, smooth lines of codes in the data stream. In the global data stream, disconnected narratives of the past – long-suppressed religious passions, suddenly resuscitated ideologies, ethnic grievances, wounded nationalist sensibility, border wars – float by without necessary pattern, liberated from time: decontextualized, deterritorialized, and dehistoricized. In the empty, cold, and weightless vacuum of the outer space that is the digital reality, there emerge now, with greater frequency, spasms from the past for a culture with pure signal as its underlying code and spectacles of the hauntological as its content.

Or is it just the opposite? If there can be today such immense technological pressure on the body, quickly eclipsing barriers between consciousness and the data stream, infiltrating feelings, facilitating the transition of subjects rising into objectivity, might this not suggest an enduring inability of the digital signal to overcome human noise? In this case, for all the relentless power of the image system, for all its carefully calibrated visual strategies for overcoming image noise – blurring the image, image subtraction, reducing contrast – what always remains is that certain sign of persistent human presence: the image noise of recalcitrant consciousness, concealed emotions, complex knots of messy subjectivity, the resurgence of complicated histories, embodied time, unresolved contexts – the "mottled, grainy textured or snowy appearance" – haunting the viral positivity of the image system. In this scenario, that which is truly hauntological – racial injustice, class struggles, gender politics – will not long be denied, just as much as the hegemonic language of the code will not long be unchallenged by the complicated realities of that which it seeks to deny: politics of the street. So, then, a future awaits of accelerating to inertia, sometimes as the triumphant sign of a culture of pure signal and at other times as the persistent presence of image noise. Falling upward into a state of cultural weightlessness, floating at escape velocity in the data stream, while remaining in a material, embodied world is the result of the collision of the radiating positivity of a culture of pure signal and image noise, that point where virtuality and materiality spin together in a society digitally illuminated by all the unreality of chiaroscuro realism. Neither pure signal nor image noise, digital culture is both at the same time. The planetary grid of an increasingly commonly shared image culture that requires constant speed of saturation can only be accomplished by seduction of image noise, those mottled, grainy, snowy imperfections – accidents, breakdowns, mutations, reversals, struggles – in the image machines that are portals of human content, as counter-gradients to the coded future of technological form.

Here, then, is our present situation. Standing midpoint between the onrushing future and the receding past, the result may be a perfect null-point of expanding cultural weightlessness, that instant when the highly energized escape velocity of technological acceleration begins to remix past, present, and future, simultaneously bringing to the surface of attention all the unresolved, hidden, still bitter contradictions of the past, the clashing social stresses of the present, and the pressure of a technologically driven future. When the speed of the imploding future slams into the gravitational weight of the past, when the complexity of contemporary history is dragged down by the past and confused by the future, simultaneously made to account for historical grievances of race, class, gender, and ethnicity while upgrading itself for life in the digital vortex, desperately seeking a pathway through the social contradictions of the present, the result is predictable – a society under the sign of *undecidability* drifting among nostalgia for the past, confusion about present, and uncertainty concerning the future. Aesthetically, this situation is perfectly symbolized by NASA's plan to crash a car-sized probe into the sun, a spacecraft plunging to its death in the fiery energies of the solar vortex. Like a contemporary recuperation in the language of science of more ancient religious practices to appease the sun god with a gift that embodies sacrificial expenditure, NASA scientists have somehow captured the spirit of inertia that is the capstone of contemporary digital society. Today, accelerating digital energies moving at hyper-speed are seemingly everywhere – creative experiments in deep machine learning to better accelerate the modification of human minds by artificial intelligence; acceleration of digital capitalism itself as it breaks free from the first phase of globalization and begins to float at the speed of virtualization beyond the material constraints of race, class, ideology, and borders most of all; accelerating digital bodies quickly learning to live at the lip of the net, finally liberated to put on the skin of information as their deepest (digital) self; and mass media accelerating beyond the spectacle of information to become random bursts of pure signal, "breaking news," and simulated feedback – all driven by the panic mood of the times with its rage without pity and hatred of existence. Under these circumstances, politics itself begins to accelerate under the sign of new fourfold trajectories – spasms of violence, drift culture, panic politics, and the resurgence of possessed individuals as emblematic signs of the times.

Technologies of the New Real explores the quickly unfolding destiny of digital reality by means of four portals to the future: DIY bodies, power under surveillance, dreaming with drones, and robots tracking across the uncanny valley. Here, the four new trajectories of social media – externalization of consciousness, synchronization of emotion, objectification of subjectivity, and virtualization of culture – provide the underlying context by which to understand the fully undecidable reality of ubiquitous surveillance, the seduction of data tracking, the spasms of violence unleashed by drone warfare, and the long march of robots towards a fusion of artificial intelligence and articulated bodies. Equally, the portals to the future represented in all their complicated power by technologies of surveillance, tracking, drones, and robots perfectly crystallize the gathering storm that is the twenty-first century with all its intimations of racialized violence, drift bodies, random episodes of panic anxiety, and spasms of individuals seemingly possessed by invisible digital codes not of their own making and leading, perhaps, to a future not necessarily of their own choosing. An exploration that is equal parts diagnostic and prognostic, *Technologies of the New Real* seeks to travel deeply into the digital future while all the while remaining attentive to the proposition that there is no code without context, no (digital) stream without obstructions, and certainly no triumphant signal without noise in the machine.

1 DIY Bodies

There is a new DIY body in town, one that might not have the cultural pedigree of the shock tattoo, the slippery word, or the enigmatic yet subtle shift of modified bodily appearance, but rather a version of the DIY body that already belongs to the future for the simple reason that it comes to us directly from a future, dreamed about, obsessed over, but not yet practically realized. Visible signs of the new DIY body are everywhere: *smart apps* that track caloric expenditure, distances walked, miles run, rhythms of sleep, of sex, of friendship, of rage, of cheating lovers lost and won; *dusty clouds of data* that rise from the travelled earth of every footstep of the DIY body as it crunches its way into some unknown database along the way; and invasive but usually undetectable *sociobots* that break the surface of the skin, all the better to gently manipulate perception, to shape imagination, and perhaps even to take up permanent residency in the wasteland of the psyche. While the DIY body to which we have long been habituated represented the lovely unpredictability of individual choice playing itself out across the surface of skin, gender, and sexuality, the new DIY body comes to us with a self that has already split: part human/part data. In fact, the body that lives in the tension of this fatal split may be the only lingering remnant of the human, since the "self" seems to have recently departed towards the gathering horizon of artificial intelligence, synthetic biology, and robotic technology – towards, that is, the larger movement of the "quantified self." When the rising city of the quantified self breaks away from the wilderness of the unquantifiable body, we can know for certain that those data clouds are also harbingers of troubles ahead for the question of human subjectivity and, with them, the eclipse of the intuitive, the ineffable, the instinctive, the numerically unintelligible but the emotionally knowable. Putting on the synthetic skin of the new DIY body with its extended sensors, creative apps, helpful prosthetics, and enabling augments is, of course, only the first step in modifying the body right out of itself in the direction of the singularity event.

Waiting for the Singularity

The streets of San Francisco are crammed these days with creative social media startups, many waiting, it seems, for technological rapture – the much-anticipated and longed-for singularity event when artificial consciousness finally undocks from human intelligence to usher in a new future of computers literally with (artificial) minds of their own and human minds as so many data points supporting the indefinite expansion of the lifespan promised by synthetic biology, nanotechnology, and artificial intelligence.

If biblical prophecies are any kind of guide, the triumph of artificial consciousness will initiate unpredictable, morphological changes of state across the fabric of space and time. The new force of ubiquitous computing may be violently rent with Big Data on one side and soon-to-be-left-behind Luddites on the other; relational processing will sweep across the land, and the body itself will finally be able to abandon its natural ties to flesh, skin, and bone in favour of the bliss of the fully quantified self.

First prophesied in the writings of Vernor Vinge, now actively promoted by Raymond Kurzweil, currently director of engineering for Google, and first given explicit social articulation by Kevin Kelly and Gary Wolf,[1] the coming of the technological singularity is at once the ecstatic promise and utopian hope of all those scientists, technologists, engineers, graphic artists, social media marketers, designers, and programmers who have dedicated their very bodily lives to the proposition that data is the new us.

Since its political inception, the theme of waiting for the messiah has long been the core eschatological trope of American society. From the first landing at Plymouth Rock by the early Puritans and the evangelical revival meetings that spread like prairie fire across the American midlands of the spirit in the nineteenth century to late twentieth-century invocations of religious visions of those to be either anointed or left behind in the days of apocalypse, the spirit of the messianic, with its troubled doubling of transcendence and despair, has long been native to American identity. Consequently, it comes as no particular surprise that, in these times, the early sunrise years of the twenty-first century, just when the dawn is lifting on the shadows of the past, Northern California is witness to the birth anew of the spirit of rapture, this time detached from previous concerns with religion and politics, and provided with a powerful digital expression in the form of technological rapture.

On the surface, the rhetoric of this latest American revival movement is delivered in the deliberately arid form of technocratic ambition – an "Internet of Things," the "quantified self," "A Data-Driven Life" – but scratch the surface of the covering rhetoric and what springs to mind are all those unmistakable signs of the spirit of rapture. Everything is there: a theology of technology driven by an overwhelming conviction that the vicissitudes of embodied experience are subordinate to digital transcendence; the will to extend life either by uploading the human mind into its AI machinic successors or by passionate faith in the born-again body of artificial DNA; the doctrine of data as a state of (code-driven) grace; and conversionary enthusiasm for the fully quantified life. While many different perspectives gather under the revival tent of technological rapture, one common thing remains: an abiding faith that technological society is quickly delivering us to a future inaugurated by a singularity event, that epochal time in which intelligent machines take command with promises of a mind merger with a data world that is fluid, mobile, relational, and indeterminate. Though sceptics standing outside the circle of technological rapture might be tempted to reduce its enthusiasm for data delirium to the larger figurations of the form of (technological) subjectivity necessary for the functioning of digital capitalism, that reduction would surely overlook the fact that the contemporary will to technology is itself driven by a more radical eschatological promise, namely that the will to data has about it the tangible scent of finally achieving what the project of science has always promised but never delivered – human relief from death, disease, and bodily decay. While Francis Bacon's emblematic treatise *Novum Organum* may have been the first to so confidently link the project of science and the heretofore quixotic quest for immortality, it was left to a contemporary techno-utopian visionary, Raymond Kurzweil (*The Singularity Is Near*), to transform Bacon's ontological ambition for science into a practical strategy for better – that is, extended – computational living.[2] In Kurzweil's promised land of blended reality, the sky is the limit, with traditional boundaries dissolving between nature and human nature, the material and the virtual, and the enduring vulnerability of the human body with its defining limits of illness and death finally transcended. Powered by advances in nanotechnology, artificial intelligence, and gene editing, the world as we have always known it will be in sudden eclipse, and all this change will take place under the rising sun of a technological ornucopia. As Kurzweil argues, "through the use of nanotechnology, we will be able to manufacture almost any physical product upon demand, world hunger and poverty will be solved, and pollution will vanish. Human existence will undergo a quantum leap in evolution. We will be able to live as long as we choose. The coming into being of such a world is, in essence, the Singularity."[3]

At first glance, this vision is only the most recent expression of the Greek concept of hubris, the cautionary tale concerning the ineluctable balance between excessive pride of purpose and mythic punishment meted out by always observant gods. Adding complexity to this reinvocation of hubris as often-fatal overconfidence, that vision of singularity is, in actuality, a doubled expression of hubris. First, there is the sense of technological overconfidence involved in breaking beyond the traditional boundaries of the specifically human in order to speak of the new epoch of "man and machine," that is, fully digitally interpolated subjects in which the specifically human merges with the extended nervous system of the cybernetic. Here, the merely human is replaced by the technologically enabled post-human as the fundamental precondition for the singularity. With the sovereign expression of technological post-humanism, the stage is set for the futurist release of all the pent-up excess of expressions of scientific determinism and technological fundamentalism that have been gathering momentum for some five centuries – transcending bodily limits, eradicating illness, ending poverty and hunger, and vanishing pollution. In its basics, this version of technological futurism, with its doubled sense of hubris and complicated alliance of recoded bodies, nanotechnology, genetic determinism, and artificial intelligence, is a creation myth – "the coming into being of such a world is, in essence, the Singularity." With techno-futurism, we are literally present at a digital rewriting of the Book of Genesis with all that is implied in terms of (re)creating the body for smoother, and perhaps safer, passage through the often-turbulent event horizon surrounding the black hole of the singularity towards which (technological) society is plunging. While the DIY body may have the "Internet of Things" as its necessary digital infrastructure and the "quantified self" as its ideal expression, what drives it forward, animating its design and inspiring its constant creativity, is, in the end as in the beginning, the spectre of the coming singularity as its core creation myth. Curiously, in the same way that Heidegger once noted that the question of technology can never ever be understood technologically – that we must travel furthest from the dwelling place of technology to discover its essence[4] – the concept of singularity, while evocative of the language of science and powered by digital devices, is something profoundly theological in its inception.

Given the sheer complexity of contemporary global society with its mixture of recidivist social movements, global climate change, fully unpredictable human desires, economic turbulence, and, of course, changing rhythms of bodily health and the many diseases of the aged and the sick, Kurzweil's vision is startling, less so for its naivety than for its feverish embrace of an approaching technological state of bliss – transcendent, teleological, and terminal. His vision is *transcendent* because its overriding faith in machine intelligence, nanotechnology, and gene research is premised on the imperative of "overcoming our frail bodies with their limitations." Here, unlike the Christian belief first articulated by St. Augustine in *De Trinitate* – with its division of the body into corruptible flesh and the perfect incorporeality of the state of grace – the newest of all the singularities is intended to lead to a new heaven of computation. It's *teleological* because this vision of the new singularity invests the will to technology with a sustaining, indeed inspiring, purpose: overcoming the unknown country of death. And it's *terminal* because it is also a philosophy of end times, certainly the end of the human species as we have known it but also the end of easily distinguishable boundaries between the "biological and the mechanical, or between physical and virtual reality."[5] In Kurzweil's vision, the nanotechnology revolution is, in fact, revolutionary precisely because it will facilitate the creative construction of new bodies and technologically enhanced minds for life beyond the current "limitations of biology."[6] It is the end, therefore, of the biological body as we have known it and the beginning of something very novel: the merger of natural biology with its surrounding environment of technologies of the post-biological – artificial intelligence, nanotechnology, molecular science, and neurobots. As to be expected, in return for the sacrifice of a natural biological cycle of life and death, the creation myth framing technological rapture has promises of its own to keep: a fully realized future of "living indefinitely" with nanobots streaming "through the bloodstream in our bodies and brains," telepathy in the form of "wireless communication from one brain to another," and improved "pattern recognition" by overcoming the inherent limitations of natural cognitive evolution in favour of "brain implants"[7] marking the inception, then triumph, of "nonbiological intelligence." In effect, the vision of technological rapture is conceived as a marvellous, ready-made (AI) toolbox for constructing DIY bodies.

When Singularity Intersects with Human Multiplicity

While singularity theory provides a highly creative, futurist account of events likely to happen when machinic intelligence surpasses the biological limits of human cognition, the reality is that singularity is less futurist than something already deeply historical. One of the key tendencies of early twenty-first-century experience is that we may already be living in the midst of the predicted turbulence and exponential rate of change associated with the singularity. With astounding advances in robotic technology, drones that are soon to be invested with ethical autonomy in making closed (cybernetic) loop decisions concerning the "disposition matrix," relentless mergers of the worlds of society, politics, and economy with artificial intelligence, genetic biology, and nanotech intrusions on the biological, the singularity – the merger of the biological and the artificial – is a decidedly contemporary phenomenon, one that is complex, intersectional, exponential, and fractured: 3D printing is capable of virtually replicating the world of material objects; research labs have announced the emergence of synthetic biology premised on artificial DNA; robotics has shed its mechanical skin in favour of taking up habitation in the neural networks of information society; and the spectre of a globalized surveillance network is made possible by the eerily animate presence of complicated systems of non-biological intelligence associated with data mining. While narrowly technocratic perspectives may like to predict the approaching dawn of a new future of singularity – with its decidedly unrealistic projections concerning new utopias of health, lifespans, wealth, and unfettered knowledge – we, the first living subjects actually present at the fateful encounter between the biological and the artificial, understand at the granular level the real-world consequences that follow the singularity. When the information blast disrupts the social, when artificial DNA effectively resequences the story of natural evolution itself, when the triumph of code works to reinforce existing inequalities in labour, business, and politics, then, at that point, we can recognize that the (technologically envisioned) singularity actually expresses itself in the language of human multiplicity.

Scenes from the Event Horizon

Life by Numbers

The American futurist Gary Wolf has argued that, until recently, self-knowledge through quantification was impossible if not "pointless." But then, in his description, things changed suddenly and decisively. Electronic sensors moved quickly from the outside of the body to its deepest interiority; mobile devices became ubiquitous; shareware culture became the norm; electronic sensors were miniaturized; and wired consciousness made its global presence known in the form of the cloud. With this, self-knowledge was ejected from the inner realm of slow contemplation into the quantified realm of self-knowledge moving with the speed and intensity of what Wolf describes as the "data-driven life."[8]

Palpable signs that we are already living in the midst of the singularity are provided by the growing cultural appeal of what has been described as the "quantified self movement." In this scenario, bodies strap on their mobile prosthetics, digitally tattoo themselves with an array of wearable electronic sensors, calibrate their social media lives by complex, flexible forms of digital self-tracking made possible by those new clouds of digital cumulus drifting across the global sky, and turn the previously unmeasured, untracked, and perhaps even unnoticed into vibrant streams of shareable data. Essentially, the surface of the body, as well as its previously private interiority, is transformed into GPS data in the greater games of augmented reality. Except, this time, data bodies are not so much using mobile phones to scan graphics that open onto a previously invisible world of graffiti, games, and advertising, but envelop the body in a big GIF (graphics interchange format) of its very own – a digital penumbra of numbers about eating, sleeping, loving, and working that provides an electronic shadow for tracking bodily activities. Suddenly, we find ourselves living in an age of the body and its digital shadow; this complex cloud of hyper-personalized data points is not just accumulated by mobile bodies as they track their way through life but is always spinning away from the body in fantastic reconfigurations of comparative databases that may be perfect receptacles for social sharing but are also measuring points for better individual living.

Thought of in purely astronomical terms, the quantified self movement is like a *protostar* – a dense concentration of "molecular clouds where stars form."[9] Here, the newly emergent data self quickly throws off qualitative cultural debris from its past, thus committing itself to the daring gamble of seeking to quantify the unquantifiable, to literally construct a DIY body, one measurement at a time, that takes close account of lessons to be learned, data to be shared, measurements to be undertaken, numbers to be calculated, results to be reflected upon, and activities to be improved, upgraded, and overcome by its digital double – *life by numbers*. In any event, for a society in which complex mergers between machine intelligence and human bodies are underway, one important adaptive response on the part of an always flexible human species is to transform subjectivity in the direction of that which is required for smooth admission to the end times of technological singularity. If the language of power is data, if the language of connection is convergence, and if the privileged value is speed, then what could be better than a coherent, comprehensive, and creative plan for reproducing a form of "self" that eerily mimics the etymological meaning of data as "thing-like"? Refusing the intuitive, throwing off the ineffable, and breaking forever with the imaginary, the quantified self movement reverses the traditional order of human subjectivity by making the thing-like character of quantifiable data both the precondition and goal of individual identity in the age of nonbiological intelligence. Unlike traditional Christian monasteries that provided physical shelter in good times and bad for the idea of the sacred and its associated religious institutions, the quantified self movement promulgates, in effect, a new order of digital monasticism that puts down roots in the psychic dimension of human subjectivity itself. With *being data* its primal act of faith, with the meticulous, even obsessive, *calculation of life's quanta* – be it empathy, happiness, sex, or cardiovascular health – as its social practice, and with *meetups* of members of the quantified self movement as its mode of confessional, this new monastic order heralds the eclipse of traditional expressions of human subjectivity and the triumphant emergence of the thing-like – the "data-driven life" as the form of (technological) self now taking flight at the dawn of the singularity.

But wait. If you were to attend one of the global quantified self meetups – and they are everywhere now – the reality is most likely the opposite. The overall thematic might be the quantified life, but what resonates is the sense of individuals trying to find themselves, perhaps puzzled by the complications of daily life, and attempting as best they can, one self-confession at a time, to put the whole thing together for themselves by talking and sharing data. For example, each participant has five to ten minutes to discuss three core predetermined questions: "What did you do? How did you do it? What did you learn?"[10] It is as if network communications are not so much about the cold indifference of relational data points but about its actual content – that whole stubbornly individual, always vulnerable, terribly anxiety-prone mass of highly individuated individuals. There is definitely a general yearning for self-improvement in the air, definitely a sense that the basic themes of Norman Vincent Peale's *The Power of Positive Thinking*, with its homage to projected self-confidence and adaptive behaviour, has escaped the power of the written text and taken up an active alliance with proponents of the quantified life. Or maybe it's something different. Perhaps talking by data is the most recent manifestation of Dale Carnegie's *How to Win Friends and Influence People*, with its insightful strategies for winning other people over to your own way of doing things by first and foremost winning yourself over to yourself.

Indeed, if one of the key characteristics of contemporary times is the seemingly relentless progression of robots towards becoming more human, it is equally the case that many humans may be in pursuit of bodies suited for better robotic living, namely the "data-driven life." In his visionary statement of life by numbers, Gary Wolf begins with the essentially theological insight that the uniquely human qualities of fragility, precariousness, and forgetfulness, while perhaps acceptable in the epoch of the pre-digital, should now rightfully be dispensed with as the original sin of the data-driven life. According to this visionary of life by numbers, "humans make errors. We make errors of fact and errors of judgment. We have blind spots in our field of vision and gaps in our stream of attention ... These weaknesses put us at a disadvantage. We make decisions with partial information. We are forced to steer by guesswork. We go with our gut. That is, some of us do. Others use data."[11] Perhaps, but then maybe Wolf hasn't read Nietzsche's *Thus Spoke Zarathustra*, with its constant refrain about the cold indifference of nature, the absolute lucidity and absolute coldness of that indifference, particularly in the face of rationally calculated human purpose. For the quantified self, data are the newest expression of nature, which just might mean that the storytelling that data evoke also has about it a very real sense of lucid indifference even in the face of human intentionality. We might want things to be different, but data reveal the real story. It is the cold eye surveying the subjective messiness of human experience, the indifferent scale of values taking calculated measure of all things, from calories burned and sleep cycles altered to the rise and fall of financial fortunes at the speed of high-frequency trading. Or is it? Maybe, in the end, what lends the austere concept of data such seductive power is less data's pure etymological meaning as the "thing-like" than something else entirely, namely that, like everything else – feelings, body images, social connections, cultural knowledge, work experience – there really is no such thing as pure data, no empty signifier floating freely outside of a complicated, dense field of intersecting relationships. In that case, when data plunge into the post-human condition, when data express their supposedly cold judgments in all those quantified self meetups, there can be such a powerful sense of yearning in the air precisely because advocates of life by numbers – whether from the tech community or not – are always complicating the numbers by private anxieties, specific intentions, and complicated feelings. That is what the confessional storytelling at all those meetups is about – not so much, in the end, life by numbers, but life itself. It is perhaps precisely in the equivocal meeting of cold data and passionate yearning, in this strange mixture of human desire to control the complexities of social experience by numbered tabulations and data's lasting indifference to the illusions of control, that we can also begin to discern future intimations of life by numbers, that we are committing ourselves anew to an approaching era of absurd data.

Tweaking Neural Circuitry

Why should the technological drive towards the "data-driven life" remain forever on the outside of the body, enabled by apps that create self-generating loops of information guiding behavioural modification? What would happen if the desire for self-tracking were finally liberated from the body's exterior surface, migrating inside the body generally and becoming fully interior to the brain specifically? What if one day the human brain could be lit up from within by means of advanced biotechnological devices that would suddenly draw into visibility that which, until now, has remained the subject of intense speculation and passionate conjecture, namely the possibility of tracking the brain's complex neural circuitry and thus potentially enabling a new era of the DIY brain – one that involves tweaking the human nervous system. An insightful report by Robert Lee Holtz titled "Mysterious Brain Circuitry Becomes Viewable" provides this comment:

> At laboratories in the U.S. and Europe, scientists are wrapping the brain in soft sheets of microscopic sensor circuits, lighting it up within using cell-sized diodes, turning it into a wireless transmitter ... [providing] a recipe for delivering all sorts of advanced technologies, such as integrated circuits down in the brain.[12]

The brain as a "wireless transmitter" or "integrated circuits down in the brain" – that seems to be a scientific prescription for a cinema of neural apocalypse in which technologies of behavioural modification move from the outside of the body to the core of its cerebral cortex. No longer, then, is there a requirement for quantified self meetups, with their contagious techno-enthusiasm for tracking metrics of all kinds. In this scenario, silent meetups of integrated circuits that are downloaded directly into the previously untrackable universe of human neurology prevail. What possibilities yet undreamed, what future still unimagined would suddenly become viable if data tracking – presently focused on that which leaves only the "faintest measureable trace" – would deliver its advanced technologies in the form of integrated circuits hardwired to the motherboard of the human brain?

The overall goal of neurological modification – actually reshaping the neural circuitry of the brain – is the essence of the DIY bodies of the future. Light up the neural circuitry of the brain, use "tiny seeds of light" to "activate networks of light-sensitive neurons," remake the brain as a "wireless transmitter," and we are instantly living in a newly emergent world of affective neuroscience: augmented intelligence, cybernetically enabled emotion, operant conditioning of neurological depression, technically facilitated happiness – a world of genetically improved senses. Neuroscientists motivated by dreams of genetically modifying the neural circuitry of the human species have already formed the usual alliance with large-scale commercial interests invested in ambitious plans to harvest neural circuitry for accelerated capital accumulation. Similar to most other spectacular digital launches, this double alliance of science and business around tweaking neural circuitry is motivated, in the first instance, by an ideology of facilitation. Who wouldn't prefer for their children, if not for themselves, the heretofore impossible utopia of neural circuitry that could be effectively modified to deliver improved intelligence, health, emotions, and physical appearance? Download integrated circuits in the brain, and human neurology would be quickly rendered the first and best of all the cognitive apps of the future, ready to practically realize the most recent advances in robotics, genetic biology, and nanotechnology. It would be as if technological rapture took possession of neural circuitry and delivered the integrated brain to the ecstasy of singularity.

However, the other side of the ideology of (neural) facilitation is the presence of integrated circuits that take command. In this sense, once neural circuitry has been lit up by those "tiny seeds of light" and once "special networks of light-sensitive neurons" have been activated and their neurological structure diagnosed, the result is likely to be brain matter dangerously overexposed and, in fact, perhaps fatally vulnerable. What and who, then, will be the DIY bodies of the future? How will issues related to class, race, ethnicity, and gender play themselves out in the approaching universe of re-engineered neural circuitry? And what happens when the previously invisible region of human neurology with "as many cells as stars in the Milky Way" abruptly moves from its sheltering darkness to the bright lights of scientific probes that want, above all, to explain the complexity of "all those millions of pathways"? From sometimes harsh historical experience, we know well that questions of visibility and invisibility are never simply reducible for their explanation to the question of technology. Who and what will be brought into visibility has always been an essentially political determination. Equally, who and what will remain cloaked in invisibility, and thus rendered exterior to traditional rights of human recognition, also involves prior political settlements concerning issues bearing on prohibition, exclusion, and disavowal. All this complexity, of course, is studiously screened away by purely technological analysis determined to finally achieve the elixir of all scientific ambition – lighting up the soft matter of the brain in order to probe its neural contents with integrated circuits. Here, the accelerating speed of technologies of (neural) facilitation easily outpaces contemporary deliberative reflections on the fate of the human nervous system first fully objectified and then harvested by the command language of affective neuroscience. William Leiss, a futurist philosopher of genomic science, once asked: "Are we ethically prepared for this?"[13] Are we ready, ethically ready, for the coming order of neural modification, with its tweaking of the human nervous system, first as a way of facilitating an improved human situation (albeit for some) and ultimately to assume full neural command of that which was previously unmeasurable, untrackable, and invisible?

Remote Mood Sensors

As if to accelerate the process of lighting up the brain and thus bring the full complexity of its neural circuitry into a greater visibility, a cutting-edge research program was announced with the aim of creating remotely controlled mood sensors, ostensibly for controlling depression and anxiety, which can be inserted directly into the brain.[14] Again, following doubled logic of facilitation and command, the ethical justification for such prototyping was made in terms of bringing urgent medical relief to traumatized soldiers suffering the long-term effects of post-traumatic stress disorder (PTSD). Given that the mood sensors are intended to be operationalized with possibilities for remote control, it might also be hypothesized that a biotechnological device of this emotional magnitude may also align itself very smoothly and without a ripple of (scientific) discontent with what the theorist Paul Virilio has described as the process of "endo-colonization,"[15] namely strategic interventions by which governments make war on their own domestic populations. As reported by Patrick Tucker in *Defense One* ("The Military Is Building Brain Chips to Treat PTSD"), the research program follows the trajectory of technologies of "deep brain stimulation." With the aim of treating extreme mood swings, including "anxiety, depression, memory loss and the symptoms associated with post-traumatic stress disorder," research teams are focused on mapping the brain's neural circuitry, with special focus on how "surges of electrical signals moving across its motor cortex express themselves in symptoms related to PTSD."[16] Brain chips, then, are being developed for modulating mood swings in subject populations – cybernetic devices injected directly into the soft tissue of the brain.

Future augmentations of the DIY body with brain chips – "invasive deep brain implants" – lend themselves most immediately to dystopian visions of mind control. Under the therapeutic cover of improving individual psychological health by reducing depression, anxiety, and mood swings, what is really being delivered to the brain is a fundamental change in the patterns of its neural circuitry. Once brain implants have been drilled down into the soft matter of the brain, the expectation is that gushers of neural data will provide new ways of mapping, then modelling, the brain's electrical networks. Once installed, brain chips could potentially reverse engineer the amygdala by changing the patterned behaviour of neural circuitry as a way of circumventing the neurological sources of traumatic injury. Once the brain has been opened up by cybernetic implants to mood-altering therapeutics, it creates the possibility of generalizing this initially purely therapeutic intervention across entire populations. In other words, it's "a crude example of what's possible with future brain-machine and cybernetic implants in the decades ahead."[17]

Perhaps, though, it's not "mind control" in the traditional sense of a political mechanics of domination but the wiring together of previously individuated brains into new forms of fused affectivity. In this case, brain chips are a two-way (neurological) street, both transmitting data to waiting sensors from deep inside the soft matter of the brain and delivering to the amygdala mood-altering therapeutics. If a future of bodies with brain chips is alarming from the perspective of received visions of mind control, perhaps that is because this project is already less a futuristic project than a deeply retrograde one. In a highly mediated culture, we have long been accustomed to what McLuhan once described as "media as massage" – electronic media that modulate the human nervous system with psychologically powerful simulacra of images, sounds, and (virtual) emotions. To some extent, inserting digital devices such as brain chips only makes obvious what may have already happened to us in that complex environment of brain/cybernetic interfaces known as the mass media. But, if that is the case, maybe what is most disturbing about brain chips for mood alteration are two of its other constitutive features. First, with this neurological experiment in "invasive deep brain implants," an ethical boundary is fatally breached, one in which the human brain is harvested as another inanimate object of vivisectioning. By creating a brain implanted with prosthetics, drilled with chip technology, carefully mapped and modelled, this project is, in essence, an experiment in rendering neural circuitry a fully alien object of radical experimentation. What is possible, then, with "future brain-machine and cybernetic implants in the decades ahead" may be a deeply ominous future in which neurological functioning is reduced to a servomechanism of more pervasive cybernetic patterns of behaviour: operant conditioning delivered by a brain chip at the speed of light optics. Second, the project involves not just brain chips as advanced expressions of wireless operant conditioning but also the construction of DIY bodies of the future built upon the triumph of the data-driven brain and the eclipse of the human mind. Here, hacking the brain by literally "jump-starting" it with electrical currents would mean that the struggle to overcome consciousness of trauma and mood swings associated with anxiety and depression would be reduced to a purely operational solution, with efforts at understanding the social origins of trauma and existential crises that may have triggered acute anxiety or severe depression eliminated from the psychic scene. Jump-starting the data-driven brain also means a big increase in the cybernetic control of human neurology and an equally big decrease in the necessarily contingent, contextual, and ineffable nature of human consciousness.

Of course, for researchers of the data-driven brain, consciousness of the ultimately consequential results of the project may well lend added visibility to fundamental ethical doubts concerning the wisdom of this latest proposal for the technological interpolation of neural circuitry. For example, if past practices hold true, the first test subjects for this experiment in brain vivisectioning are likely to be animals involuntarily sequestered in laboratories, then perhaps even selected groups of army veterans who may be told that participation in this experiment aimed at implanting cybernetic sensors into the brain is a precondition for continued medical treatment. Equally, if mood swings are to be placed under remote (medical) control, what is to prevent the dark side of data – viral contagions, aggressive hackers, stolen or misplaced memory sticks, broken codes – from being introduced quickly and decisively into the deepest recesses of the soft matter of the brain? Sharper (brain) images result, then, but also blurred ethical vision.

Twisted Memories

Our country is in chaos. But it's a great day to be an American.
                                                    – CNN, 4 July 2020

There is a new (scientific) game in town involving neural modification of the cerebral cortex. It is called optogenetics, and its stated aim is to "build a window into the brain" with the creation of biocompatible neural implants, making possible, for the first time, direct, unmediated communication between the digital world and the human brain. In other words, better neural implants for injecting the power of algorithms directly into previously unknown, unexplored, and uncharted brain matter. With this modification, the final boundary between bodies and machines will be dissolved, fully exposing the soft matter of human consciousness to modulation by digital algorithms.[18]

Fittingly enough, for a technological society seeking to accelerate at the speed of light, optogenetics promises to marshal the power of light to provide a powerful gateway to neural modification of the brain. The optogenetics project is funded by DARPA (Defense Advanced Research Projects Agency) as part of its long-term "Brain Initiative"; the promotional literature focuses on the deep affinity between technological innovation and scientific expressions of the morally good will, which, in the case of optogenetics, involves hopeful scientific attestations concerning medical advances potentially actuated by manipulating individual brain neurons with powerful three-dimensional light patterns. In this case, blast artificial sunshine onto individual brain cells, and the hoped-for medical outcome will be something akin to neurological magic, with the anticipated gains for optogenetics said to include neurological treatments for ailments ranging from the physical (impaired vision) to the psychic (mood disorders).[19]

That is the hopeful rhetoric anyway. But dig a little deeper in the mission statement issued by DARPA, and in the small print towards the bottom of the project description, there is an intriguing mention of the potential for optogenetics in the future to implant "sensory percepts" in human consciousness. It's not just calming down neurological patterns of behaviour associated with mood disorders by replacing pharmaceuticals with light-based treatments but something much more politically significant, literally using a future version of biocompatible neural implants to inject new visual memories into the brain. With this technology, an intriguing link is established between neuroscience and contemporary cinema, where the artificial production of manipulated memories has long been the focus of the cinematic imaginary, including the basic conceits for *Total Recall, Memento, Inception, Dark City,* and *Blade Runner.* Like the prevailing model of evidence-based scientific knowledge itself, which is premised on an innovative combination of pioneering theoretical speculation in search of real-world experimental verification, optogenetics promises, in effect, to provide a real-world demonstration of what until now has been the exclusive territory of cinematic visions of the future, namely to insert twisted memories directly into the neurostructure of the cerebral cortex. While in its preparatory research phases, zebrafish larvae and mice will be the testing ground of choice for optogenetics, the expectation is that the Brain Initiative involved in linking light-activated optogenetic proteins with generative neurons will quickly climb the evolutionary scale to summit in the tethering of human consciousness with neural implants.

Sound dystopian? It promises an unfolding technological future powered up by optogenetics where, with neural implant devices yet to be created, the brain will be opened up to radical psychic surgery by light-actuated technologies with the aim of disappearing lived memory, substituting in its place artificial memories – "sensory percepts" – of a past that will probably be all the more vivid, compelling, calming, soothing, and haunting precisely because it never actually happened. However, in the usual way of things technological, if optogenetics as a dystopian gateway into neural modification of the brain can sound so ready-made menacing, that is probably because it is a latecomer to the real-live games of politics in the age of neural modification by those other circulating light-actuated screens of mass media. Here, producing false memory syndrome, memories of events that are all the more real because they never actually occurred, is the normal business of contemporary media imaginaries. The skilful combination of spectacular televised spectacles, mass suggestion, and mesmerizing rhetoric provide an instant gateway into the politically effective neurological modification of cerebral cortices living under the personal and collective stress of impotent anger, unabated rage, and hysterical anxiety. For example, consider the implantation of synthetic percepts in the form of false memory syndrome so spectacularly performed on that always fabled day of moral reckoning in American political consciousness, 4 July 2020, where, with memories of other Independence Days from Sherman's military conquest of Atlanta to the Declaration of Independence itself, President Trump took to the symbolically hallowed, and deeply colonized, ground of Mount Rushmore in the Black Hills of North Dakota to rework American political memory with silence on the pandemic, mockery of Black Lives Matter, and the declaration of a new civil war based on warring racial relationships and featuring a beleaguered and threatened population of vulnerable white Americans assaulted by menacing hordes – Black Lives Matter and their allies in the streets as part of a resurgent "left-wing fascism" – coming to steal American history, to obliterate cultural memory itself. Everything was there: televised spectacle with the carved presidential faces on Mount Rushmore in the background, a mass crowd as a real-life stand-in for the hoped-for mass audience of American television viewers, and mesmerizing rhetoric of the war spirit based on equal mixtures of panic fear and proud belligerence that was immediately circulated everywhere through the media stream. This contemporary inversion of the Sermon on the Mount, with its emphasis this time on the war spirit not the love spirit, did not have to wait for future breakthroughs in the neuroscience but deployed light-based technologies ready at hand, the screen culture of television and social media, as its very own neural implants for communicating new memories – true for some, false for others – from the digital world to the only psychic structure that really counts politically, the liquid, fluid brain matter of American mass society, in order to accomplish the neurological modification of the collective

American cerebral cortex. That the neural implant was successful could be immediately verified by those experimental demonstrations in the streets: white jaguars being driven into crowds of protesters and shootings of anti-racism protesters by palpably anxious white men and women armed with guns locked and loaded.

In contemporary quantum culture, there are no necessary causal links, only deeply entangled relationships among fast-moving particles moving in the same web of space and time: positional, relative, and connected. This vector of twisted memories, as its moves at high velocity through the mediascape, sometimes takes the form of a futurist announcement from DARPA concerning engineering neural implants for fast communication between digital reality and the brain; at other times, it's streamed in all its hauntingly prophetic vividness by the cinematic imaginary and, then again, makes its presence felt affectively as the basic code of contemporary political neurology. Here, neuroscience, futurist cinema, and political campaigning are themselves different twists taken in this entangled story of twisted memory: recombinant in its movements, bendable in time and space, and always travelling at high velocity. Of course, as with all important technological changes, neuroscientific research in optogenetics remains at the periphery of human attention until, once fully realized, the consequential results are moved instantly from the edges of human awareness to its very centre. Perhaps more than is realized, optogenetics promises to be the capstone of technology's relentless movement from the exterior of the body to its deepest interiority, involving as it does the stated engineering aim of providing a gateway into the cerebral cortex, not only changing behaviour but providing for the potential implanting of new visual memories. Following the ideological formula that eases public accommodation to radical technological change, optogenetics was introduced in the language of facilitation, namely the promise of facilitating medical breakthrough. Once successfully adopted, the real implications of optogenetics as command language of the approaching age of neural modification will be imposed: neural implants for remembering a past that never existed, algorithms for mood modification, three-dimensional light-based treatments for screening away the differences. The once and future migration of neuroscience from the language of facilitation to command will itself be facilitated by the fact that we are now preparing to enter a future we have already long visually inhabited through the cinematic imaginary and which, moreover, we have already experienced at a deep affective level through the fabulations of twisted memory as the emotional alphabet of contemporary politics.

When Synthetic Biology Rides the Wave

We are actually transitioning from a *Homo sapiens* into a *Homo evolutis* – a creature that begins to directly and deliberately engineer evolution to its own design.[20]

Conditions are perfect for surfing in La Jolla, California – sunny sky, steady breeze, and gigantic waves finally finding their way to the Pacific shoreline, swelling up to beautiful crests just before the whole (wave) scene dissolves again and again into a bone yard of broken patterns of water ebbing onto the beach. On this particular morning, there are dozens of surfers riding that magical California edge of bright sun and killer waves, some just bodysurfing but most trying to find the sweet spot of those cresting waves, that momentary physics of the barrel where bodily balance, fast motion, and the curve of the cresting wave exists for the millisecond that is the take-home measure of the perfect wave. Now all that is pushed to the (pleasant) background of my attention as my mind is locked in deep, reading Greg Bear's prophetic book *Blood Music* in a beachfront café located just steps away from the Scripps Institution of Oceanography, with its fabled marine research of life related to the watery element of the physical universe. In this whole scene, a lot of surfing is going down. Certainly, there are those incredible surfers of the waves just offshore, but there are also those marine biologists engaged in a kind of intellectual surfing of their own, this time trying to ride the waves of those sometimes perfect patterns of watery life forms. There's also some serious surfing taking place in *Blood Music*, although this time it's not human bodies tracking cresting waves or marine biologists looking to catch and ride the edge of insightful findings, but a story concerning the future of nanotechnology: a science fiction fable of artificial cells that have escaped the lab, taken possession of the body of a graduate researcher, and then literally surfed the biological material of that single body until those artificial cells propagate beyond synthetically infected flesh to change the physiological structure of the entire environment. Aesthetically, the image of the future offered by *Blood Music*, with its story of artificial life and computation come alive, is similar to those eerie images painted by the surrealist artist Max Ernst, where human bodies, inanimate objects, vital animals, and mythological symbols bind together into a common morphology. Politically, it's anticipatory of Bill Joy's warning that, while a computer crash might mean the inconvenience of some lost data, crashing the basic codes of life runs the danger of taking down entire environments, if not suddenly terminating the natural evolution of the human species.[21]

In the usual way of always incommensurable thought, my mind might have the apocalyptic futurism of *Blood Music* in its foreground and those

scenes of rhythmic surfers in its background, but my situational awareness is short circuited by a news alert from my always on mobile that
transmits a news release from Scripps Research, located in La Jolla,
California, where it is announced that researchers have developed a
bacterium "whose genetic material includes a pair of DNA 'letters' or
bases, not found in nature." In other words, evolution is suddenly on
hyper-speed, where the basic genetic alphabet of life with its two pairs of
DNA bases, A-T and C-G, is suddenly ramped up by bio-engineering to include a "third, unnatural pair of bases." The leader of the research team
draws the wondrous, yet potentially ominous conclusion: "This shows
that other solutions to storing information are possible and, of course,
takes us closer to an expanded DNA biology that will have many exciting
applications – from new medicine to new kinds of nanotechnology."[22]

While the news release was enthusiastic in its account of synthetic
biology delivering on its promise of a new alphabet of life, my own
"exciting application" of the development of artificial DNA was tempered
by the immediate thought that, try as I might, I could not sequester in the
background of my perceptual field: Was it really possible that, only three
decades after the dystopian fable traced by *Blood Music*, events first written as literature have leaped the divisional boundaries of fact and fiction
and become the modelling principle for the future of the real? Is *Blood
Music* the skin of the new real of synthetic biology and artificial DNA?

There can be little equivocation with the claim that synthetic biology, with its transformative creation of artificial DNA, is the future of
the DIY body first, and perhaps later even of the DIY planet. Brushing
aside the seemingly feverish efforts by neuroscientists to stake proprietary claims on rewiring cognitive networks, whether by drugs, tracking,
or implanted cyber hooks, synthetic biology has introduced the fundamental game changer of artificial life. For example, while contemporary social and political thought continues to debate the contentious
relationship between power and life – whether, that is, power speaks in
the name of (normative) life or in the more disciplinary name of death –
synthetic biology envisions something entirely different, specifically the
creation of previously unimagined forms of artificial life, from synthetic
cells to the artificially constructed bodies of soldiers, astronauts, and
workers that take full advantage of "an expanded DNA biology." More
than "life by numbers," the "quantified self," or "remote mood sensors,"
and going beyond mechanistic images of the re-engineered brain as a
"wireless transmitter" or an "integrated circuit" with neurons to be lit up
and neural pathways to be "jump-started," synthetic biology provides a
dramatically new creative principle – artificial DNA. Here, the addition
of a "third, unnatural pair of bases" to genetic history does not simply

promise "solutions to storing information" or expanding DNA biology but introduces a fundamental element of uncertainty into the living world. While injecting a freewheeling and essentially designer note of the recombinant, the unnatural, and the artificial to the biological process of coding "life on earth" will undoubtedly facilitate many novel and worthwhile applications, it also means taking final possession of the question of life itself. Consequently, when genomic scientists envision multidisciplinary approaches linking together molecular biology, chemistry, computer science, and electrical engineering, what they are really articulating is the gateway to the future – a gateway to enhanced possibilities for "assembl[ing] biological tools to redesign the living world."[23]

At this point, thinking at the intersection of ocean-driven scenes of California surfers and science fiction hauntologies of *Blood Music*, I wondered if the unnatural world to come will also someday experience for itself those strange and enigmatic fractures of broken meanings, uncomfortable fits, and clashing cosmologies of the heart and mind that seem so unique to the human species about to be left behind. Measured by the first, truly global burst of excitement that greeted the Scripps announcement – an excitement less, to be sure, about the foregrounded text of a novel scientific breakthrough than what seems be the really existent, animating subtext, namely that we are speaking openly and positively about redesigning molecular building blocks for the "living world" – and judging solely by the positive response to this drop-dead end of evolution, end of (natural) story press release, there is an unqualified smoothness to the future of artificial DNA. While artificial DNA might not, as synthetic biologists like to claim, be allowed to escape the laboratory, that does not preclude active experimentation with synthetic DNA in the many other laboratories of power and capital – weaponizing synthetic biology, creating highly specialized artificial life forms to maximize capital accumulation as well as minimize labour unrest, technologically enabled, eugenic dreams of synthesizing the "perfect child." No longer the "terrorism of the code" in any particularly negative sense, it's a future scripted in all its smoothness, transparency, and perfectibility by the rising (genomic) signs of synthetic biology.

Yet, for all that, there is still a lingering sense that, in the future, even the most artificial of all the artificial DNA will come to recognize that the mythic fate of the artificial – the ancient art of artifice – is always necessarily doubled. Certainly, every artifice first expresses itself in the language of perfect simulation – a smooth coding of the living world by biological tools that only work to enhance "exciting applications." But, of course, the secret of all the great masters of the art of artifice is the hard-won realization that what motivates the artificial, what really renders believability to the theatre of artifice is precisely the intangible elements of undecidability, imperfection, and, indeed, latent error that is always carefully masked by the staging of the artifice. In this case, as in (natural) life, so too in (artificial) life, the fact that every fully accomplished perfect surf ride ends in the boneyards of just another wave on the beach might just intimate that the future logic of synthetic biology already contains its own boneyards, that what presently remains unsynthesized, unthought, and unconsidered is the ghost-rider in the shadows of artificial DNA. Could it be that resuscitating something of the spirit of the human, that which is presently policed away by the totalizing logic of synthetic biology, is the once and future destiny of artificial DNA? Or perhaps the reverse is true. If *Blood Music* is the skin of synthetic biology, swarms of mutating cells, like nature before them, will be indifferent to human fate. That would mean the future of synthetic biology will probably cast natural indifference against human artifice as its likely fate. In this case, we are in the presence of new (molecular) building blocks for a very traditional story.

Remember the unanticipated, premature death of Dolly, the first of all the android sheep that, for all its artificial resuscitation by the scientific hubris of genetic engineering, could not escape its fatal destiny of accelerated, synthetically enabled aging. Just as we can acknowledge with some confidence that every massive wave is doomed to crash and every breakdown can be a potential breakthrough, so too even the science of artifice can never really escape that messy tangle of mythic destiny, complex ambitions, complicated dreams of the sub-real, and utopian dreams of transhumanism that is the continuing singularity event of the new real. In this case, the future of synthetic biology, with its creative breakouts of artificial DNA, nanotechnology, and fabricated xeno-organisms, remains fully uncertain in advance – fully undecidable, that is, until that future moment when the synthetic imagination actually begins to ride the wave of unsynthesized reality onto the beach of life itself.

Technologies of Suspended Animation

Following Heidegger's fateful insight that understanding the essence of the question concerning technology is never far away but always close at hand, never, that is, hidden away in mythic stories of secret origins but something always proximate to the post-human condition, what signs can be deciphered? What lessons can be drawn from these scenes taken from the event horizon: synthetic biology riding the wave, life by numbers, tweaking neural circuitry, and remote mood sensors? On the surface, these are discrete stories from the data-driven life, whether expressed in all its subjective enthusiasm by the quantified self movement or by technologies specializing in re-engineered neural circuitry, invasive brain implants, and biological experiments in developing artificial life forms as radically new pathways for a literally post-human evolution. Again, following Heidegger, it may be the contemporary human fate to be caught in the way of a larger technological destiny – its foundations, morphology, and ultimate direction, all of which remain unclear – although its transitional momentum is felt clearly and decisively at every historical turn. Indeed, several generations after Heidegger's reflections "On the Question Concerning Technology,"[24] the revolution in technological affairs that his thought was both attentive to and prescient about has seemingly solidified its grasp on contemporary societies with such dynamic and apparently unstoppable power that we can actually begin to discern the overall trajectory, if not the terminal destiny, of the will to technology. Again, the destiny of technology lies closest to us: for example, stories of the quantified self as raw data unfold to tell a story, a ribbon of fact, a narrow path of what is promised to be transcendence. Sometimes the unexpected comes to call: a blip, a pause, a catastrophic rupture, or, perhaps, just a broken line of code. Or, again, it might be stories from synthetic biology about the development of an approaching epoch of "biological superintelligence" – artificial life forms constructed specifically to carry forward into a still unknown future the complicated collusion of humans and machines at the speed of algorithmic processing, with the bodies of articulated robots, artificial orifices of synthetic senses, and the planetary skin of the "Internet of Everything." Viewed retrospectively, the latter is how Cisco, the California futurist telecom of things related to wireless networks, routers, and network-switching mechanisms, typically preferred to describe the accelerated, seemingly hyper-exponential rate of change associated with the internet. While the internet may have begun in the late twentieth century as a visionary, yet relatively limited, communicative order, by the early years (2008) of the twenty-first century it had already generated its own wireless offspring – the "Internet of Things" – accompanied by the inevitable tech euphoria enthusiastically describing the "unstoppable path" of Big Data applications, from cloud computing and 3D printing to mobile payments, as ways of accelerating the wired future – linking, tracking, monitoring, and amplifying life in network society.[25]

Always speaking with the confidence of a cartographer about a digital future of which it is itself one of the key communicative architects, Cisco no sooner announced the advent of the Internet of Things than only several years later – in a Schumpeterian-inspired act of creative destruction – promptly abandoned the latter conceptualization in favour of an even more grandiose vision: the "Internet of Everything." Here, what is privileged is the power of *connections*, not things, not smart devices but smart connectivity where, in the fully connected world of the "Internet of Everything" with its "billions of things, trillions of connections," connectivity is the key, not just simply linking bodies, devices, invisible data infrastructures, and things but actually making "dumb things" smarter, facilitating the routines of daily life, empowering the world with its home-based digital assistants, automated cars, caring devices, and robots mingling in the here and now of everyday life.[26]

It's a perfect expression, then, of the rapture of technological connectionism, with its layering of the language of "smartness" onto the otherwise inert world of routers, switchers, and interfaces and its hijacking of the therapeutics of "helpfulness" on behalf of the "trillions of connections" among "people, process, data, and things."

The Quantified Fetus

And why not a quantified fetus? Digital euphoria of this order produces many helpful results that illustrate possibilities for connecting "everything" in unexpected ways. For example, a company called Bellabeat provides both a digital device and an app to serve as a fetal monitor, providing continuous, real-time, biologically sensitive read-outs of the baby's heartbeat. It can also track baby kicks and even the mother's weight gain.[27] Expanding beyond the traditional intimacy of a mother's intuitive feelings of care and concern for the baby growing inside her, Bellabeat emits heartbeats as digital sound bites that can be shared with family and friends over the internet – literally the Internet of Everything, including downloadable, shareable, real-time heartbeats of an unborn baby. The quantified fetus can be heard anywhere, anytime, through the wondrous "power of connections." In an uncanny intimation of the real time of the digital superseding the biological time of the human, digital histories of fetal activity, as in the case of Bellabeat, make possible digital life histories spanning a longer time continuum that the chronological life cycle of humans that begins, at least in the West, with actual birth. Promoted as a digital device that can be trusted, Bellabeat, of course, may also have the unintended effect of undermining the mother's trust in her own intuitive feelings for the invisible, yet emotionally palpable, presence of her unborn baby. A curious case ensues, then, of increased digital sensitivity based on real-time data concerning the health of babies and a soft, yet insistent, undermining of a mother's actual emotionally based feelings for the well-being of her unborn baby. It's not so much the old question concerning which to trust more – machine read-outs or intuitive, inchoate feelings – but something else. In this case, does the power of (digital) connections also have the power to deliver us to a world of (emotional) misconnections?

"Turning the Body into a Password"

But why stop with the quantified fetus when it soon will be possible to inhabit a DIY body that is password protected? Google's Motorola "skunkworks" division prototyped a new digestible digital device (Motorola's Edible Password Pill) that, once swallowed, would instantly transform the human body into an authentication tool for accessing digital domains from smartphones and laptops to digitally swiped doors, whether offices, garages, or homes. Nominated by *Time* as one of the "twenty-five best inventions" of the year it was launched, the accompanying description included the following: "Swallowed once daily, the pill consists of a tiny chip that uses the acid in your stomach to power it on. Once activated, it emits a specific 18-bit EKG-like signal that can be detected by your phone or computer, essentially *turning the body into a password.*"[28]

Following the overall logic of technological incorporation, where data increasingly breaks the skin barrier, moving from its outer surface to its biological interiority, this digital device upgrades the body with the power of actually becoming its own interface, merging the "power of connections" to data flesh with such biotechnical seamlessness that the digitally authenticated body smoothly and effortlessly merges with an Internet of Everything. Regina Dugan, former head of DARPA, former leader of Google's advanced technology team, and now chief executive officer of Wellcome Leap, has remarked about the body as its own "authentication token": "Once swallowed, it means that my arms are like wires, my hands are like alligator clips – when I touch my phone, my computer, my door, my car, I'm authenticated in. It's my first super power. I want that."[29]

Working from the perspective that "electronics are boxy and rigid, and humans are curvy and soft,"[30] Google's aim is to complete the always difficult last few millimetres of connecting the "curvy and soft" flesh of until now organic human beings with the geometric grid of digital connectivity. Consequently, it promises a future of modulated technology – soft, ubiquitous, pliable, smooth – sometimes camouflaged as electronic tattoos on infants (data tracking for better security) as "authentication tokens" in the supposedly hyper-cool style of e-tattoos or as "stretchable circuits" for detecting concussions in sports injuries. Unconsciously adopting the language of mimesis, this form of body invasion by the contemporary generation of data snatchers, from the Motorola Edible Password Pill to digitally coded rap tattoos, is brilliantly disguised as a biological appendage – subtle technology with the added benefit of conferring "super power" on fully authenticated bodies.

Between Life and Death

Perhaps the quantified self has already moved on to the diagrammed body, that point where digital devices are so deeply embedded in our psyches – from quantified fetuses to password-protected bodies – that technology has now become a read-out of the human life cycle. When what should properly be on the periphery of human attention becomes central to perception, neurology, moods, or the human nervous system itself, there is bound to be some damage. It is not so much that, under the pressure of technological change, the human sensorium has now been turned inside out, resulting in radically split human senses – partially still interior to individuated bodily histories and partially circulating at the speed of digital circuitry – but that there is a growing prohibition against self-awareness of what has been lost with the appearance of the diagrammed body. There is no digital device that does not leave a bodily trace, no fusion with a synthetic life form – whether a net bot with an inflated sense of artificial intelligence, a supposedly "smart" form of machine-to-machine communication, or a cyber implant in the theatre of synthetic biology – that does not revise memories, disrupt feelings, disappear the precious singularity of that which is not only unique but ineffable: the relationship between a mother and her suddenly data-driven baby, bodies viewed as inauthentic because they are unauthenticated, or life itself filled with jagged edges, slow trudges, and always messy confusions of being a life form of organic matter in an increasingly dematerialized world.

Consequently, while we can be aware that the "power of connections" is swiftly delivering us to a future capable of producing quantified fetuses and password-protected bodies, what remains unclear is the ultimate cultural, and perhaps even existential, impact of the triumph of the transhuman. Considered in terms other than dystopia or utopia, is it possible that such adventures in transhumanism – powered by visions of technological rapture and the singularity event, practically implemented by the quantified self movement, and replete with experiments in vivisectioning neural circuitry by synthetic biologists – are fundamentally changing the meaning of life and death for the human species as a whole? Perhaps it's not a future of technological rapture but an indefinite period of suspended animation in which the human species, as a life form kept waiting for the singularity event that may or may not ever arrive, makes its final, feverish preparations for a fateful crossing-over point between machines and humans, yet not wanting to be untethered from digital prosthetics and definitely not anticipating that very real crossing-over point – the always solitary experience of death – without helpful technologies wrapping themselves around the "soft and curvy" matter of the body organic as it terminates.

There is a revealing report in the *New Scientist* about a new emergency technique in suspended animation ("Gunshot Victims to Be Suspended between Life and Death")[31] that bears directly on larger issues related to technology, culture, and life itself. The story recounts how surgeons at a Pittsburgh hospital are now experimenting in suspended animation for victims of traumatic injuries – by guns, knives, or blunt objects – as a way of stopping blood loss, thus gaining bodily time in order that their lives can later be saved by the necessary medical interventions. One surgeon is quoted as saying: "We are suspending life, but we don't like to call it that because it sounds like science fiction. So we call it emergency preservation and resuscitation."[32] The technological procedure used in this trial is straightforward: once the aorta has been clamped, a solution of saline is pumped "through the heart and up to the brain," and the patient's temperature is reduced, with the result that "at this point they will have no blood in their body, no breathing, and no brain activity. They will be clinically dead."[33] Hopefully, though, they are not dead for long, since after the necessary surgical interventions, blood is flushed through the body, the saline solution purged, and the patient's body warmed up by its own circulating blood, with this (redemptive) medical conclusion: "We've always assumed you can't bring back the dead. But it's a matter of when you pickle the cells."[34]

Now, while this story is an intriguing one concerning the truly liminal boundaries between life and death, it may also be a preliminary glimpse of the fate of the human species generally and the DIY body specifically, as it is flushed with a saline solution of synthetic technologies, its key organs clamped shut with password-protected apps, its body temperature definitely cooled down by increasingly antiseptic loops of cold code, and its neural circuitry placed in a state of suspended animation waiting for resuscitation by technological rapture. While medicine, like all of science before it, cannot in the end overcome the finality of human mortality, the greater ambition of contemporary technology, particularly in its transhumanist expression, is captured perfectly by the surgeon's insight into the decidability of previously undecidable matters of life and death: "It's a matter of when you pickle the cells."[35]

2 Power under Surveillance, Capitalism under Suspicion

Surveillance Never Sleeps

Surveillance never sleeps because it lives off data trackers designed to never forget. Algorithms have become cabinets of digital memories with sensors that attach themselves to the words we speak, places we see, and even thoughts not yet expressed. Our lies and truths lived through our nights and days.

Like the sleeplessness of data itself, always mobile, circulating, and recombinant, network surveillance lives under the strict obligation to police the full circumference of digital being: all those financial algorithms rendering instant, real-time judgments on questions of economic solvency; algorithms in the form of technologies of "deep packet inspection" for supervising violations of civil rights; algorithms for economic espionage in the name of national security; algorithms for pleasure, for gaming, for better apps; algorithms for tracking, recording, and archiving the habitual activities and errant breaches of any human heart that makes up life in the data torrent today. While at one time insomnia referred only to a human sleep disorder, now a new form of insomnia – *data insomnia* – has been created.

We can therefore examine some reports from the field of a pervasive machinery of surveillance that never seems to sleep, with its data farms, archive terror, face printing, embedded sensors, smart bodies, and cold data. But first, however, we'll look at some surveillance reports from the real world of contemporary politics, where mass surveillance without saving vision is increasingly challenged by counter-surveillance from citizen journalists giving (video) witness, with the result that the faceless surveillance world of vast data farms, hidden sensors, and cold data is brought into sharp focus with instant streaming by citizen journalists in the city streets intent on bringing to the surface of visibility, and perhaps political accountability, hard truths that are spoken by power to its previously silenced subjects. While surveillance that never sleeps has quietly and pervasively been set in place by technological platforms that are the skin, nerve tissue, and organs of the digital world, all perfectly emblematic of the triumph of a technologically enabled future of the post-human, counter-surveillance by citizen journalists with mobiles and video cameras testifies to the animating strength of bodies in the streets, bodies that generate, time and again, compelling scenes about what happens when power can no longer rely on a silent web of digital devices, embedded sensors, and pattern recognition algorithms moving fast but is forced to announce its presence in brutalist terms with militarized police, flash bangs, and charging batons. This fatal contest between the silent machinery of mass surveillance without vision and counter-surveillance as giving witness to social injustice is, I believe, the really existent context for understanding the politics and consequences of surveillance that never sleeps. In this case, perhaps we are now transiting from a society under the sign of surveillance with its corollary in "surveillance capitalism" to something very different, namely power under surveillance, capitalism under suspicion, and mass media distrusted by an increasingly dissenting population.[1]

Surveillance Reports

Mass Surveillance without Vision/Counter-Surveillance as Giving Witness

D.C. had no problems last night. Many arrests. Great job done by all. Overwhelming force. Domination.
> – @realDonaldTrump, 2 June 2020

The Nazi slogan for destroying us ... was "Divide and Conquer."
> – former defense secretary Jim Mattis

When the symptomatic signs of fascism have displayed themselves, triumphantly and without a hint of apology, at the highest levels of American government – the redefinition of American cities as a "battlespace"; the denunciation of peaceful protesters as terrorists; the militarization of police forces; government surveillance planes overhead capturing cell phone data of protesters; the imposition of curfews as part of a generalized state of siege; frequent occurrences of racialized violence and active, violent suppression of civic dissent by batons, tear gas, smoke canisters, rubber bullets, pepper balls, flash bangs, helicopter, horses, and military vehicles – then the question of surveillance becomes a storm centre of contemporary politics. Here, the role of surveillance as a powerful tool of the national security state in identifying, tracking, archiving, and isolating sources of democratic dissent quickly moves from the background to the suddenly crystallized foreground of the gathering debate concerning surveillance. If the self-proclaimed aim of this newly articulated version of fascism is to establish "dominance" over the population by the use of "overwhelming force" supplemented by the militarization of policing, then a necessary tool in accelerating domination will be the generation of ubiquitous forms of surveillance that would permit the state to identify not only existent sources of popular dissent but also to anticipate, and thus suppress, the future of dissent itself. Today, surveillance never sleeps because power increasingly thinks of its own domestic population in terms of a hostile insurgency controllable only by "overwhelming force" and "domination." Of course, not cowed by a surveillance apparatus acting at the behest of "overwhelming force," witnesses to scenes of police violence, protesters subjected to "domination," and observers of human rights violations create a vibrant, critically engaged network of counter-surveillance by mobile cameras that is as ubiquitous in its instant diffusion across the mass media as it is effective in mobilizing efforts towards political change. All the while, in this raging scene of mass surveillance and popular counter-surveillance, the triumph of visual media culture, with its predictable rhetoric, carefully managed narratives, and framed images, is instantly undercut by the return of oral culture. As the political theorist David Cook notes, "Trump's use of Twitter is a form of oral culture that mesmerizes the individual in short aphorisms, slogans and quick spells reducing book culture which is too slow to compete. A culture that produces the vicious turtle."[2]

That the present is a dangerous moment in the decline of the United States from an ongoing experiment in somehow reconciling representative democracy at home and hegemonic imperialism abroad to domestic fascism is intensified by the fusion of four crises appearing simultaneously in American public life: an *epidemiological* crisis under the viral sign of COVID-19; an accelerating *economic* crisis taking the form of mass unemployment and precarious wage-labour aggravated by the pandemic; a *social* crisis focused on racialized violence and the deployment of militarized policing; and a *cultural* crisis in which the stakes are high, namely the life and death question of whether or not the future of contemporary fascism will take root in public opinion and, consequently, in the deep emotions that form private subjectivity. Certainly, overstressed by crises from without and depressed internally by feelings of anxiety, panic, melancholy, and frustration, it is only one small step for individual subjectivity to seek shelter in the comforting rhetoric of charismatic political leaders who strike the words of overwhelming force and domination on the anvil of a bitterly divided population. Or could it be the opposite? Perhaps it is not an inexorable slide into an era of reactionary politics enforced by mass surveillance and energized by a pervasive silent majority of white privilege; perhaps something very different may arise, namely a fundamental social rethinking of police violence, racialized power with its apparatus of control focusing on mass surveillance, in the direction of social justice, first in the streets and then in the homes, the minds, and the workplaces of a growing majority disenchanted with the status quo. Videos of racialized violence, then, become possible gateways to insurgencies of social justice.

Trapped in the Bubble of Technology

What happens when the surveillance apparatus of the national security state is met with counter-surveillance by its citizens? What takes place when police violence, so often directed against people of colour, is matched by a waiting network of real-time video cameras by people, young and old, willing to be citizen journalists, willing to give witness? And what is the future of "surveillance capitalism" itself when citizens, increasingly aware of the granular intensity of network surveillance, begin to go off-grid, campaign for stricter regulation of large technological platforms, or dream of a reinvention of the internet itself as a publicly accessible global forum of communication in which issues of personal privacy are sacrosanct? What happens, that is, when surveillance by algorithm, coding, and reverse-engineered screens and mobile communications is finally brought out of the darkness of the secrecy of technocratic proprietary knowledge into the bright illumination of critical public debate? When surveillance is met with counter-surveillance, when tracking by machine is matched with counter-tracking by citizen journalists giving witness, what is the future of the real world of surveillance?

From a strictly political perspective, surveillance technologies are effectively designed to function as faithful ideological representations of the society from which they emerge and on behalf of which they are tasked with protecting it from unexpected attacks, unplanned intrusions, unanticipated threats, and potential new dangers. Consequently, if a society such as the United States – the self-proclaimed leader of the "free world" and the self-appointed moral guardian of "exceptionalism" – is both the leading model of hyper-capitalism as well as the financial and political spearhead of an aggressive, hegemonic global empire, its surveillance technologies will necessarily be modelled on the preservation of a dynamic form of power, the interests of which are truly global in their spatial extension and complicated in their interaction with all the boundary conditions of politics and commerce operating in real time. In this scenario, surveillance needs to constantly scale up its technological ability to track, by data harvesting, screens as cameras, and consumer household devices, an always tumultuous world perceived by the masters of hegemonic power as permanently insecure, perpetually replete with unexpected threats, always repopulated by newly emergent political oppositions. Ironically, with its heavy emphasis on the surveillance of spatial extension, bringing the globe under the microscope of advanced technologies, mass surveillance is often blindsided by the slow moving waves of time itself. For all its harvesting of massive data flows and its algorithms for pattern recognition, mass surveillance, trapped in the bubble of technology, often experiences difficulty in detecting those subtle intersections at the heart of the real world, those unpredictable collisions of racialized violence, economic inequities, systemic racism, rising hostility to asylum seekers and refugees on the part of white nationalists, sudden surfacing of ancient ethnic hostilities, insurgent religious fundamentalisms, and counter-ideologies forming at the edges, and sometimes at the very centre, of empire. Of course, if surveillance cannot escape the bubble of technology, if, that is, surveillance technology cannot detect subtle, but decisive, changes in history as time, then it is probably doomed in advance to give witness to the future of its own empire in ruins. One sure and certain indicator that contemporary surveillance technology has already entered its terminal state is the contemporary scene of American politics, where the fabled "homeland" is suddenly perceived by those in power as ideologically precarious, with citizens feared as potentially insurgent, and where the American republic itself is reframed as a military "battlespace." In this circumstance, surveillance technologies previously justified as responses to potential threats from outside the borders of the homeland turn inward, quickly repurposed as powerful instruments of "overwhelming force" and "domination." What was, in the recent past, a tacit state of affairs – a very real, but unspoken alliance between the national security state and technological platforms – now surfaces as the necessary condition for linking the algorithms of "surveillance capitalism" with political control of the American battlespace. Here, the machinery of surveillance is the tip of the spear of contemporary politics. Alexa recording domestic conversations, browsers acting as digital shadows, ubiquitous surveillance cameras in the streets, mobile devices with their geo-tracking of travel and location, automatic reading of licence plates, planes flying over protests in Lafayette Square near the White House collecting digital data from the smartphones of the citizen dissenters below – all these take on a very different meaning when not viewed simply in terms of capitalist accumulation but of the politics of total domination with all the potential deleterious political consequences that that implies.

However, in the usual course of things, bubbles are always susceptible to being burst. That is the importance of populist counter-surveillance today. Large, global, energetic, and passionately idealistic social movements that struggle to move the arc of justice forward are often volatilized by the pervasive dissemination of images depicting the brutality of power, ecological disaster, Indigenous insurgencies, asylum seekers dying by drowning, indefinite incarceration, and sometimes by indifferent neglect. From Hong Kong to the cities of Europe, from real-time Indigenous videos of heavily armed RCMP detachments destroying ancestral Indigenous encampments in Canada to the streaming of videos of the murder of young Black males in the United States by gun, cars, and knees on necks, counter-surveillance is how the real world of time enters the media stream. Here, the slow and patient gathering of personal information by all the networks of the digital world is matched by the fast and impatient visual interpretation of the use and abuse of power in the politics of the streets of the digital world. While mass surveillance is about data compilation and heightened political control, counter-surveillance by citizen journalists concerns visual truth-telling of scenes of social injustice. The former analyses massive flows of data for pattern recognition; the latter turns the time-based waves of social life, sometimes disruptive or smooth but always in motion, into ways of listening, seeing, hearing what is actually going on at the jagged edges of racial inequality, class struggle, gender discrimination, political injustice, and exclusion from human rights based on moral recognition, whether by ethnicity, race, religion, or geography.

SpaceX and a Burning Police Station

Consider, for example, the deep irony of that media day in late May 2020 when televised scenes of a burning police station in Minneapolis, Minnesota, surrounded by angry, grieving night-time crowds of protesting citizens illuminated against a fiery background contrasted so sharply with that other media event taking place at exactly the same time. Here, SpaceX's launching of a manned space vessel for a mission of replenishment to the International Space Station was also the key media event of that tragic day. This time, though, the scene was not burning cities but fiery space launches, not angry Black Lives Matter witnesses on the streets but breathless technological hosannas from media commentaries about SpaceX, and most certainly not federal troops and National

Guards with the presidential words of authorization embedded in their minds – "when the looting starts, the shooting starts" – but a resplendently arrayed demonstration of American space force power prominently highlighted by President Trump himself. The empire might be in decline, streets may be in riot, citizens of colour may have dissented against the policing of white supremacy, but at this SpaceX moment in the history of the United States, the illusion of American hegemony in space, if not in the actual earth, was ready for a triumphant blast off, not to the stars of deep space but to their orbiting token, the International Space Station.

But what happens when the smooth narrative of this deep-space moment is broken by counter-surveillance, this time not with a streamed video image or a cell phone shot but counter-surveillance by the power of imagination, by establishing unspoken connections between the clashing events of that turbulent media day in late May? What happens, that is, when counter-surveillance is liberated from technological devices, becoming instead a new way of critically seeing the world: making connections between data feeds; bringing into visibility the illegible remains of racial, economic, and political inequalities; giving voice through imagination to the silenced, the excluded, the prohibited? In this case, if the SpaceX launch can be such a compelling display of the technological imaginary of a society permeated with the logic of racial hierarchy, then what is the meaning of these contrasting media images: bursts of flames burning with the anger of riots in the night, and spectral space launches by day? Are we witnessing a fatal struggle between metaphor (SpaceX) and metonymy (torched police station) in the contemporary American mind? And if, in fact, the essential symbolic meaning of the SpaceX launch was that it was a highly visible reminder that the animating energies of America's empire of technology – the spirit of instrumental activism and the magic of a society infinitely adaptable to changing economic circumstances – was about to be exported to the stars, what symbolism was to be had in the burning remainders of that day, in the charred embers of the police station and the angry crowds of protesters, not projected towards a future in the stars but present in the remains of the earthly night?

We therefore have two sharply clashing media images: one about astronauts travelling to a space station in the sky, the other depicting those gathered in grief over a police murder in the streets. One image is the imaginative technological creation of Elon Musk, celebrated by American media for his enthusiastic spearheading of the power of positive (technological) thinking – a model of the much-cherished business myth of the entrepreneurial spirit, an evangelist of technological futurism who has managed to successfully launch himself beyond the bitter racial politics of South Africa to the more friendly, experimental, and supposedly racially antiseptic digital terrain of west coast America; the other is the product of the persistence of that most fundamental of all clashes in the heart of contemporary America – rage against racialized violence and the very real persistence of an ideology of white supremacy launched historically from the ruins of the Civil War to take root seemingly everywhere in the language of the everyday: in jobs, housing, health, security, prisons, and policing. Are we witnessing here the rise of the uncanny in American thought: dreams of technological redemption from above, revolt against racialized violence from below?

Surveillance technology closes its eyes of digital perception when actually confronted with the complicated intersections of social history. There is no digital cloud, no information flow, no pattern analysis, no determinate data, no harvested archive of behavioural histories that would enable understanding of the strange paradox of days of rage in the streets and technological utopia in all the launch pads of deep-space futurism. But if the apparatus of mass surveillance is bereft of the saving power of vision, it does not mean there is not much to be learned from that sad day in May when the real world of racialized social reality crossed paths with dreams of technological transcendence. When mass surveillance falls silent concerning the really existent drivers of the unexpected, the unpredictable, the unsecured – unchecked aggression in police uniform; the persistence of white male rage; the contemporary political reality of panic, fear, anxiety, and melancholy; the stubborn continuation of patterns of racial inequality, class disparity, and gender discrimination – it is fated to be marginalized as a predictor of the future, as an intimation, that is, of the likely consequences of a society where technology moves at the speed of escape velocity while the lives of Black people and people of colour remain stuck in the inertial weight of social oppression.

Consequently, two very different futures of surveillance are presented: one, the shared product of the national security state and corporate technological platforms that consistently "see" the world only through digital algorithms that are deterritorialized, decontextualized, and dehistorized; the other, the insurgent creation of citizen journalists committed to the deeply ethical practice of giving moral witness to social injustice – a form of critically engaged media practice that has the effect of providing images of a world that power would prefer to prohibit from witness, an invisible world of the silenced, excluded, oppressed, and abused. While the former is about control, the latter concerns taking back the streets of imagination, the alleyways of life lived in the full spirit of multiplicity. While mass surveillance reduces society to the codes of digital inscription, the algorithmically knowable, and the data detectable, counter-surveillance is always about the immeasurable, the off-grid undetectable, the strangeness of the unknown, the immense, radiating power, creativity, and courage of those living under the punishing sign of social invisibility. Two worlds materialize: fast data and slow bodies; technologically enabled mass surveillance and humanly enacted counter-surveillance; monitored bodies treated as data trash and bodies struggling in the streets of power, with brutal streaming video evidence of the real meaning of abuse value everywhere. A future is envisioned of data archives, quantum surveillance, embedded sensors, and the ever-expanding network of the "Five Eyes" in a contemporary political situation that intimates so forebodingly that this future is also tangled up with very real violence. Having witnessed so often and in so many places the deployment of surveillance technologies for purposes of overwhelming state power, we find the issues that haunt any discussion of surveillance today concern the ultimate uses of information gathered, the alliance between data and domination, often resulting in unaccountable policing, whether of the imagination or bodies – the targeting of peaceful protesters, increased mass incarceration, and pre-emptive sabotage of democratic dissent. Hence, we have capitalism under surveillance, power under suspicion, and media distrusted as the essence of a future now shaded by the clashing horizons of counter-surveillance by citizen activists and mass surveillance by always watching Big Brother algorithms.

Data Insomnia

Lightning Storms in the Data Farm

A story running on the wires in 2013 reported another shutdown in a major National Security Agency (NSA) data centre. While there weren't "arcs and fires anymore," experts on site hadn't yet diagnosed the problem, noting that "they [had] figured out how to prevent lightning, though."[3] Now finally powered up and fully online, the NSA Utah Data Center (codenamed Bumblehive) is a prodigious tower of (digital) babel in the beautiful mountainous terrain of Bluffdale, Utah. Not far from the now vanished site of Fort Douglas, which was originally constructed to defend older lines of American continental communication including the stage coach line on the Oregon Trail and telegram facilities, the spyware data centre also occupies a curious intersection between theology and technology, situated as it is in a community that *Wired* magazine describes as the largest Mormon-based polygamist community in the United States. Reportedly occupying 1.5 million square feet and costing over one billion dollars, the Utah Data Center is, in effect, the electronic cerebral cortex of a vast data harvesting system aimed primarily at gathering foreign signal intelligence but also at "harvesting emails, phone records, text messages and other electronic data."[4] As described by James Bamford in *Wired*, this project is one of "immense secrecy," migrating intercepted data from satellites and cables through cables and routers to vast data archives "including the complete contents of private emails, cell phone calls and Google searches, as well as all sorts of personal data trails."[5] It's literally a global memory machine absorbing, recording, and databasing all the sights and sounds of digital Babylon that is life today.

Definitely anticipating an unlimited future of information accumulation, the Utah Data Center is described as capable of storing five *zettabytes* of data, sufficient storage space, that is, for the next hundred years.[6] Of course, while technically astounding in its ability to harvest in one isolated Utah data site the world's global communications potentially spanning an entire century, NSA's spyware centre has experienced very real problems at the double interface of unpredictable nature and human political ingenuity. While lightning storms crackled across the otherwise austere architecture of its massive data servers, and "arcs and fires" seemed to break out spontaneously at the merest hint of data flowing, those human beings waiting to be harvested of their "patterns of life" were themselves engaged in creative forms of cyber-protest at the very gates of the data kingdom. Petitions were presented to the local town council demanding that vital water supplies to the data centre be immediately terminated. Since cooling water is a critical requirement for any data farm that plans to mine the data skies with all the power enabled by five zettabytes of storage memory, this local protest in defence of civil liberties had the potential effect of eclipsing secret cryptography in favour of ground-truthing by water. With natural protests taking the form of fire in the air and human resistance privileging blockages in the flow of water, it was almost as if all the mythic furies associated with the four fundamental elements of the universe – fire, water, air, and earth – had suddenly assembled in the face of this monument to absolute (digital) knowledge. As for earth, protests involving this classical element constituted a literal read-out of the meaning of grounded resistance. A group of local civil libertarians adopted a highway running in front of the Utah Data Center for the sole purpose of holding up protest signs to passing motorists, while all the while engaging in the good citizenship practice of actually tidying up the highway and its immediate environment. Consequently, a curious case emerges of a supposedly frictionless NSA data centre smoothly aggregating complex streams of global data while buffeted by very real lines of friction involving lightning in the (data) sky, dammed up water, protesters disguised as highway cleaning crews, and very strange encounters in the desert air between the dream vectors of technocracy and polygamy. Perhaps what is really happening here is something that is not captured in all those data rushes of foreign, or for that matter, domestic signals of intelligence and information awareness, namely that, like all previous demands for meaning, this urge to absolute knowledge of the human universe is always quickly outrun by the complex particulars of humanity and nature alike. Not really as Camus supposed mythic indifference to the demand for absolute meaning to be, it's something more subtle in its appearance, specifically that the technological dream of a perfect spyware universe of frictionless flows of information always generates in its wake unexpected and fully unpredictable lines of friction. That those lines of friction have no possibility of easy absorption into the hygienic and closed universe of tens of thousands of humming data servers does them no dishonour. It would simply indicate that a universe predicated on the security to be provided by the terrorism of the code is probably already doomed one hundred years in advance by that which cannot be avowed, included, or permitted – whether local citizenry gathering to turn off the cooling taps of water, lightning flashes across the server horizon, or that greatest line of friction of all, the loud media sound of Edward Snowden as he adds his own line of friction to the data harvest.

It's not just lines of friction, though. There are also strange, and deeply enigmatic, symmetries. What the NSA is constructing in a desert bowl near Salt Lake City is a genealogical record of the digital future, capturing one batch of signals intelligence at a time until that point over the next one hundred years when some bright cryptographer in a still indeterminate future time will find in all that Big Data not simply patterns of life but patterns of whole societies, of republics, democracies, empires, tracing the rise and fall of sometimes clashing civilizations, contested visions of political economy. It's literally a history of the future traced out in the data stream. However, what is truly enigmatic in its implications is the present-day fact that there is not simply one, but two, major experiments in patient, genealogical research underway in Utah. Certainly, there is the technological probe of the digital future that is the NSA's Utah Data Center, but there is also the theological tracking of family genealogy that has long been underway in nuclear-bomb-proof caves near Salt Lake City. So, then, we have two genealogies, the first aimed at cryptographic analysis of information culture – past, present, and future; and the second a theological archiving of detailed family genealogies, tracing its arc from the ancient past to the still unknown future. Of course, it could just be coincidence, an unnoticed fact of surveillance history, that the theologically signified region of Salt Lake City with its famous history of the Mormon Trek led by Joseph Smith from the darkly wooded valley of Sharon, Vermont, to the promised land of Salt Lake City was chosen by the NSA as the site of its first major server farm. Perhaps the choice of Utah as a spyware centre was made for the usual pragmatic budgetary reason: cheap electricity, plentiful water, with the added advantage of sometimes hiring, according to some media reports,[7] Mormon missionaries as NSA data analysts, given their acquired skills in foreign languages. It could also be the case, though, that this convergence of technocratic and religious interests in the question of genealogy, one present and future oriented and the other privileging the past, is symptomatic of a deeper convergence between technology and theology on the question of absolute knowledge. In this case, what might be actually happening in the scrublands of Utah are two deeply iconic exercises in truth-seeking: the one dealing with hidden patterns of signifiers in all those flows of global information and the other focused on equally hidden patterns of theological truth-saying revealed in the genealogy of family histories. While this convergence might make of the NSA only the most recent manifestation of what could be described as the redemption quest akin to a form of the new Mormonism in American public affairs, it would also make of Mormon theology, with its tripartite focus on redemptive visions, missionary practice, and patient genealogy, the premonitory consciousness of the animating historical vision of the NSA itself. The link would not be crude, in the sense of an open, avowed affiliation between the NSA and Mormonism, but something more subtle and thus more deeply entangled: namely that the NSA as the self-avowed secret spearhead of cybernetically sophisticated technological adventurism also has its eyes on the prize of merging the redemption story that is the essence of the "American dream" with (digital) missionary consciousness and cryptographic genealogy. In other words, it would be an unfolding symmetry of the Book of Mormon and Big Data in the mountains, deserts, and salt lakes of all the Utahs of the data mind.

Fast Surveillance: K-pop and Whiteout Wednesday

I'm at the inaugural K-pop concert in Toronto and just loving every moment. Going against the cultural grain, it's not convened at a cool music emporium in the evening hours when moon magic and the spirits of darkness so beautifully refract the spirit of song, dance, and enchanted musicians, but in that most unmusical of places, the Toronto Convention Centre, in the dead centre of business downtown and in bright morning sunshine, starting office worker early and closing before the

evening hours really start to stir. Now, I had read all the grumbling cri-
tiques of K-pop, this intentionally market-streamed, manufactured series
of designer bands emerging from South Korea and, like other popular cultural phenomena before it – manufactured cars, cosmetics, and
fashions – sweeping the globe with a gathering, mesmerizing energy, its
mostly young fans, teens and pre-teens, united by the power of social
media in intense discussions of what's up in the always shifting, always
changing world of K-pop. And it didn't help that the business architecture of the Toronto Convention Centre lived up to its name – concrete
drab, cave-like vast, with lots of dead space overhead just weighing down
any signs of life from the crowds below. But nothing prepared me for the
magic of K-pop and the animating, charismatic energy of its thousands
of young fans that day. Four stages were set up, some for fabulous performances of dance, song, and intricately choreographed movements
by bands as a whole and sometimes by breakout individual performers;
and some for instruction in the flesh liturgy of the K-pop performance:
stage-worthy dance instructors giving lessons in K-pop bodily moves; cos-
meticians with accompanying tools of the trade – mascara brushes, glitter, eyebrow tweezers, lipstick lush – providing lessons concerning how
to create the K-pop performance body. As one cosmetic instructor said
to appreciative fans, "you can never have enough glitter for a great K-pop
performance." On the side, vendors were selling K-pop merchandise directly out of suitcases, with CDs disguised to look just like best-seller dystopian novels, shirts with band names, and everywhere the sounds and
sights of the sprawling K-pop empire of passion and delight. Here, the
traditional distinction between performer and audience was reversed,
with the audience very much a powerful performance element in its own
right. Young dancing bodies broke out of the audience at K-pop dance
lessons to display brilliant dance improvisations of their own creation;
fan faces everywhere matched the chill cool of their bodily appearances
to professional instructions from the stage; and just everywhere K-pop
performances were instantly paralleled, and sometimes preceded, by the
singing and dancing of fans who, from long and deep embeddedness
in social media, knew every story, every band drama, every song, every
move. I thought I was attending a K-pop concert in Toronto, but what I
was really experiencing was the musical Woodstock of a newly rising generation of the young, the peaceful, the loving, with a communal feeling
just so strong, so passionate in its idealism for dreams of a better world.
As my granddaughter Claire Surat said when I mentioned before the
concert that K-pop had the cynical feel of a manufactured product, "well,
doesn't it depend on what they are manufacturing?" She was absolutely
correct in her lucid intelligence. After twelve hours in K-pop heaven,

with thousands of fans screaming, fabulous dancing, and so peaceful, so beautiful enactments of the power of band performances – Kard, The Rose, Zion/P, Veri Veri – I dropped my set critique and went over to the side of the enchanted, the hopeful, the dreamers. Later in the year, it came as no surprise to me that K-pop's global community of young fans made a truly decisive intervention in the anti-racist rallies that swept the world after the police murder of George Floyd – not by tokens of the moralized good will, but by flooding the hashtags "White Lives Matter" and "Whiteout Wednesday" to shut down flows of racist twitters.[8] When the leading K-pop band, BTS, announced that it was donating one million dollars to anti-racism causes, young fans immediately took up the challenge and matched this donation by crowd-source funding in twenty-four hours.

Here, then, is the question: What's surveillance in this new age of the energizing idealism of K-pop fans and the racist tweets of "Whiteout Wednesday"? How do you track, record, and archive the beautiful rhythms of a global fan base congregating for an instant in crowding out racist hashtags? And, for that matter, how do you do surveillance of a hating heart? What happens when K-pop fans are doing their own surveillance, moving swiftly and seamlessly between K-pop drama on stage and creatively activist politics in the (digital) streets? How does mass surveillance keep up with the speed of counter-surveillance? How does it predict the unpredictable? How does it anticipate the spontaneous, the immediate, the decisive? How does surveillance capture the paradox of global fan loyalty to a style of Korean music performance that has been described as "manufacturing humans" to those very same fans demonstrating by their political activism the most profoundly humane of all qualities – empathy, critical engagement, storming the digital barricades of racist hashtags?

If contemporary technological society is a prototype of the future integrating individuals into a machine-readable world of AI, quantum computing, and deep learning, then that future will surely be one saturated with surveillance – soft, vaporous like a data cloud, omnipresent, operating at the periphery of attention, everywhere facilitating our integration into a machine-readable world but also sometimes menacing, probing, tracking, evaluative, perfectly probabilistic – remembering our digital past and anticipating pathways we are likely to take on the digital road ahead. This future, the future of a technological society intent on securing the final links between bodies and machines, will be one of surveillance moving very, very fast. Everything today is preparatory to that future. In a world of fast surveillance, surveillance practices – political, economic, social, and cultural – will form the omnipresent background noise of life in the wires – ubiquitous, smooth, sometimes helpful, at other times sinister – functioning like a flickering signifier that seeks to mould itself to digital behaviour, to machine read the varieties of digital experience in order to facilitate opportunities but also to steer the digital experience in certain directions. The prime marker of all information societies of the future, fast surveillance is information moving at such speed and intensity that it becomes the electronic cloud that surrounds us, seduces us, facilitates our every action, and evaluates our behaviour, always archiving, remembering, managing us, and using algorithms for making probability judgments on which way our perception and behaviour are likely to spin in the future.

Perfectly transversal, moving at the speed and granular intensity of deep learning, artificial intelligence, and quantum computing, fast surveillance always exists in two opposite states at the same time – noise and signal, wave and particle, facilitating creative opportunities, taking command of the direction of digital life. In the new world of fast surveillance, we become the elementary matter of surveillance – human qubits – floating in the cloud of surveillance that surrounds us. Never stationary, the cloud of surveillance has spin, like an electron around an atom – sometimes centripetal operating at the periphery of attention, at other times centrifugal like a surveillance state under the sign of Orwell's dystopian vision of the future – patiently recording our every action, carefully parsing the meaning of our past behaviour for probability judgments on our digital choices for the future, always prompting us to pay attention, carefully sorting us into the threatening and non-threatening, lurker and creative participant, citizens loyal and not loyal, customers satisfied or dissatisfied, consumers attentive or indifferent.

Like a flickering signifier, fast surveillance is never reducible to one of its functions. It assumes many digital appearances, whether the brilliant digital galaxy of TikTok that is driven forward to machine reading our future through the prism of AI or the more menacing presence of the massive databases of the national security state. Wired pleasure *and* digital dystopia, then, are the future of fast surveillance.

There is no longer digital innocence in the unfolding future of fast surveillance. We know by long immersion in the real world of digital reality that we are always being tracked, monitored, archived, and evaluated – prodded, probed, and seduced – and, in response, we have sometimes become brilliant avatars of new counter-surveillance strategies. Consider the messianic, all-consuming interest of the rising digital generation of pre-teens and teenagers who have already opted out of the traditional role of passive recipients of mass media, becoming instead active, creative, deeply absorbed participants in the new digital worlds of TikTok,

Snapchat, YouTube, and Instagram, fascinated in the past by the image repertoire of Vine, knowing Snapchat and Instagram by (digital) heart, always pressuring the system to move faster, produce more novelty in imaging, provide new video production breakthroughs, create magical apps for linking the multiplicity of imaginations that we all are, develop technological platforms that match the speed of digital desire. Here, all the smartphones and play stations of social media are no longer simply digital devices for facilitating communication but creative gateways to improved digital perception. And sometimes, just sometimes, as in the case of K-pop fans, better digital perception combines with an abiding, affective desire for social justice, instantly remaking digital streaming into politics of the street resurgent on the net.

In response, powered by AI, motivated by dreams of deep learning, and enabled by algorithms focused on understanding patterns of consumer interest, the image system is forced to run at hyper-speed, responding in real time to the fickleness of human perception, compelled to support a frictionless environment, always anxious about unexpected drops in human attention, quickly panicked by instant changes of state whereby rapt attention to the entertainment apparatus suddenly and with no warning shifts suddenly into absolute indifference. In this version of the digital cloud that is fast surveillance, surveillance tracks, absorbs, stores, and sells data accumulated from the global communities of users to a waiting web of online advertisers, while all the databases of the national security state relentlessly acquire information about digital subjects that will prove indispensable whenever power decides to pounce. But, in this ongoing game of seduction – challenge and counter-challenge – the emerging digital generation practices counter-surveillance of the system as well, making demands for better performance, drifting in and out of attention, sometimes going off-grid as a way of escaping the information dragnet, always read to spin away instantly, restlessly shifting from one previously popular digital gateway to another. To the system's compulsion for technological perfection, the new digital generation responds only with greater bursts of perceptual creativity, brilliant fickleness, inspiring capriciousness, and critically engaged media immersion. To the system's demand for technological order, algorithmic predictability, and deep learning, the new digital generation sometimes responds by suddenly going dark, being unpredictable, digitally agnostic, always drifting, dumping one app for another without cautionary warnings and certainly without displays of gratitude for past digital services. To the system's political demand for loyalty monitored by all the very best of the hidden smartphone cameras, microphones, and data archiving that the national security state can afford, the digital generation has a sure-fire response, namely that surveillance, fast or slow, can never detect the bitterness of the human heart, hear the anger of a body always offline, and thus undetectable, to its online simulacrum, and that, in any event, patterns of past digital behaviour are definitely never predictive of the complicated desires, complex behaviours, and contradictory choices of consciousness moving at the speed of light. Fast surveillance is always forced to play catch-up with the digital drift of slow imagination, slow desire, and slow bodies. Fast surveillance is always undermined by slow life, by life moving at the slow speed of earthly experience framed by the unmarked markers, the uncoded signs of race, class, gender, ethnicity, and sexuality.

Confronted by the fickleness of desire, the unpredictability of imagination, and the intransigence of the individual human spirit, the gamers of surveillance attempt to better position the system for the future by going quantum, by making surveillance an all-enveloping cloud that surrounds us, that becomes a hyper-reality in itself, absorbing massive streams of data, moving at high velocity, and always flickering, sometimes wearing the mask of an entertainment app, at other times attaching itself to the ear, eye, and mouth in the form of a smartphone, and at still other times playing the role of the all-watching, all-monitoring hectoring superego of the national security state.

Contrary to received opinion that feared the age of quantum surveillance would mean the end of individual privacy, the opposite has already turned out to be the case. From one perspective, the surveillance system has always experienced great difficulty in making sense of individual behaviour for the simple reason that past action may not necessarily be predictive of future behaviour, that data mining knows nothing of the rich interiority of human passions, that data is not intention, and that people reveal themselves digitally as they usually do in the flesh: beautiful mixtures of chaos and order, complicated wishes, and neon dreams, black and white choices with lots of ambivalence and second guessing in between. Besides, the crisis of vulnerability that sticks like digital glue to all contemporary technological systems is that the latter only function in data space and time: not the embedded world of lived time but the hyper-world of virtual time; not materially embodied space but the hyper-space of digital reality. Outside the overcoded, overproduced, hyperventilated, speeded up world of technological perfectionism, outside the perfectly hygienic, algorithmically coordinated, machine-readable world of AI, deep learning, and quantum computing, the technological system is absolutely vulnerable in its blindness. It cannot hear the complicated sounds of hidden, obscure human intentions. It cannot see the motivating anxieties of human passion. It cannot understand the rise of anger, resentment, fear, rage, and revenge-seeking as the great existential crises of the twenty-first century. Even within the closed, wired, shuttered world of digital perfectionism, the algorithms are noteworthy for their strange mixture of technical smarts and emotional dullness. For example, consider the challenged behaviour of Netflix algorithms as they try to anticipate and accurately predict what films users might like based on past choices, not realizing the magnitudes of capriciousness, fickleness, and creativity contained in the question of choice where, in fact, what makes humans human is that they are not usually content with being repeat-repeat machines; they will always work to subvert themselves by being digitally perverse, unpredictable, unframed, uncoded, with consciousness often energized by unconsciousness of the self.

In an embodied, material world, where we often live the future as past with the present as a specious moment thrown in between, and where curiosity typically trumps repetition and the emotional swirl of passion and desire always take precedence over hygienic logic, what's a digital algorithm to do? While it is the master of a system of surveillance that has already broken beyond the classical model of physics and gone quantum in terms of fast information processing, the technological system is itself drifting further and further away from the enduring complexities and complications of the human condition. In an epoch where questions of race, class, religion, gender, and ethnicity have become, in very real time, the existential cloud of the world we inhabit, sometimes peacefully but increasingly with spasms of bitter political divisions, religious zealotry, and incipient class struggle, the surveillance system has literally nothing to say. It performs analytics, not complicated understanding. It runs the numbers, not the politics of the street. It profiles without interpretation, tracks without understanding, archives without complexity, records without seeing, stores information without knowledge. Fascinated with all the pathos, ruins, and wasted data of the past, it usually has nothing to say about the splendours and terror of the future. Outside the closed field of data, fast surveillance is emotionally dysfunctional. It follows digital footprints and sometimes even manages to translate those footprints into larger patterns of movement by all the digital tribes of the world, but it is oblivious to face-to-face conversations, since those who seek privacy increasingly go offline, as much as it is myopic on the question of concealed human intentions and torn-up human hearts. The system may run fast at the speed of subatomic particles, machine readability, and algorithmic logic, but its fatal flaw as a surveillance system is that it only works effectively in the data field. Slip the codes, unlink from data, go off-grid, turn attention to that conflicted, turbulent, invested world that is the individual human life or even the life of a nation, clan, tribe, empire, or caliphate, and the system of surveillance suddenly runs on empty. It requires human addiction to life in the codes to be effective. It literally needs life outside the wires to be eliminated in order to function successfully as an effective surveillance system. Which is why, of course, surveillance now rushes headlong at the speed of quantum computing towards an electronic cloud that surrounds us, saturates us, and does not let in any social sunshine, racial distress, or existential rain. In order to function smoothly, fast surveillance ultimately requires the impossible – machine-readable populations, scrubbed-down hygienic subjectivity carefully sequestered within hardened data borderlands, with an army of menacing drones for the remainder of the powerless, disavowed, excluded populations outside the new digital cities, nations, and empires of the future. Against this dream of technological perfectibility stands, of course, the counter-evidence of K-pop fans and "Whiteout Wednesday" as a very visible, very readable sign of brilliant unpredictability of human empathy, the fabulous unreadability of K-pop fans taking down racist hashtags by (digital) day and night.

The Chinese Algorithm

Social Credit and the Socially Discredited

If China's national ambition of surpassing American empire as the key driver of twenty-first-century commerce and power is ultimately successful, what are the implications for the global future of its ubiquitous surveillance practices? Alexandra Ma reports in *The Independent*: "The Chinese state is setting up a vast ranking system that will monitor the behaviour of its enormous population, and rank them all based on their 'social credit' or as it was first introduced for financial transactions – Sesame Credit. The Great Wall of China has now been replaced digitally by the Great Firewall."[9] While the social credit system justifies intense, intimate, and daily surveillance practices in the name of social "trustworthiness," state surveillance also follows harsh political prescriptions. As Lily Kuo and Helen Davidson comment in *The Guardian*, there are many examples of political activists denied access to technological platforms, including "three prominent human rights activists [who] have accused Zoom of disrupting or shutting down their accounts because they were linked to events to mark the anniversary of the 1989 Tiananmen massacre or were to discuss China's measures to exert control over Hong Kong," as well as activists associated with the "Hong Kong Alliance," who reported their Zoom accounts blocked to prevent discussion of an extradition bill "that caused mass anti-government protests in Hong Kong."[10]

China may not yet be the future empire of its dreams, but it is most definitely a world leader in intrusive surveillance practices. With the aim of promoting "good citizenship," the Chinese government has announced its intention to rank every Chinese citizen by their social credit standing. Implemented by tech platforms already possessing vast archives of personal data, as well as city councils governing large populations, the social credit system is envisioned as a futurist model of social engineering, with rewards and punishments related to social behaviour. What constitutes "good citizenship" is, of course, formulated by the Chinese Communist Party. In what seems to be a vast experiment in social eugenics, the behaviour of every Chinese citizen will be closely monitored for deviations from prescribed behaviour, with every deviation from Chinese norms for "civility" immediately resulting in additions to or subtractions from the social credit standing of citizens. While rewards include perks in faster travel, hotel bookings, access to better education, medicine, and restaurants, punishments involve being banned from travelling by train or planes, throttled internet service, sudden blockages in access to these necessities of life, and, most importantly, being added to a publicly distributed blacklist as a form of shaming for "bad behaviour."[11] Consequently, the Chinese government is planning ubiquitous state surveillance, where the destiny of individual life itself is measured on the collective scales of social credit and the socially discredited, the included and the excluded, the authorized and the prohibited.

While current forms of surveillance in the West often involve governments secretly conspiring with leading tech platforms to download data generated by every citizen, the experiment in Chinese social media dispenses with the secrecy, choosing instead to openly social engineer desirable social behaviour in its citizens. Here, the power of Chinese technology is marshalled on behalf of a political experiment in social eugenics, one aimed at cultivating social behaviour that would fit within the borders of the new Chinese citizen for the twenty-first century. Contrary to media expectations in the West that social credit is the epitome of Orwell's dystopian vision of the future, it appears that social credit is immensely popular in China. Moving quickly from the outside to the inside of society, social credit has rapidly achieved reputational status as a sure and certain indicator of a person's social standing in society. A high social credit score is to be touted to the sky; a low social credit score is a new form of social shaming. In effect, an intrusive system of state surveillance has rapidly become a universal exchange principle for measuring everyone's social standing. If contemporary technology is driven forward by a doubled logic of *facilitation* (better communication, education, entertainment) and *command*, then this vision of social credit takes the ideology of technology to its extreme, making the doubled language of facilitation and command the everyday, just-in-time language of travel plans, hotel reservations, train and plane tickets, expanded opportunities in education, possibilities in medicine, entertainment, and jobs. In this very active, granular, and intense futurist model of operant conditioning for purposes of social engineering the behaviour of Chinese citizens, ubiquitous mass surveillance slides smoothly into social eugenics.

Or is it the opposite? If there can be such massive commitment on the part of power in China to bend the direction of social behaviour by a full-scale commitment to operant conditioning with its meticulous system of rewards and punishments, is it because the real world of contemporary China has nothing to do with social order but rather with a chaotic flow of multiple powers, clashing interests, unrestrained individualism, creative imaginations, and unchecked domination? Social credit, then, is proposed as a veneer of social order, engineered into existence by ubiquitous surveillance and driven forward by a model of technological perfectionism, covering up an underlying reality of social chaos – an economy requiring constant state stimulation to maintain its growth rate; political order maintained by a clear separation between the unitary and forcibly unchallenged power of the Communist Party of China and the highly creative economic entrepreneurialism of the population; social cohesion based on the strict policing and systemic oppression of all ethnic dissent; and ideological order maintained by a celebration of nationalism, itself always haunted by the fear of China quickly slipping back into bitter social divisions, ethnic rivalries, and the turbulence of warlord culture. Like all previous closed orders of socially managed behaviour, the dreamland of social eugenics is vulnerable to the slightest hint of social dissent, ethnic struggle, creative individualism, or politics in the streets. Policed by digital data, managed by carefully constructed archives of personal behavioural profiles, and inspired by the ideal of a society exhibiting perfect social integration, the system tends to entropy without necessary injections of creative energy from below generated by chaotic flows of political dissent, clashing interests, intersecting ideals, multiple identities, clashing faiths, and competing destinies. When the unitary digital universe of social credit touches the third rail of actually existent, actively suppressed difference – social, political, religious, ethnic, economic, ideological – it is immediately undermined, only capable of preserving the managed order of social credit by more and more open applications of disciplinary power. For example, witness the inability of the Chinese government to successfully contain active dissent in the streets of Hong Kong, the lasting historical significance of the events in Tiananmen Square as the spectre of insurgent political opposition that haunts China's political leadership, the show trials of political opponents within the shadow world of the governing party of the empire, the mass incarceration and equally "political re-education" of the populations of Uzbekistan and Tibet, or the very real political rage directed by Chinese state media against the practitioners of Falun Gong, domestically and internationally. Confronted by the reality of really existent social change, social eugenics can only constantly inflate its behavioural data ambitions, expand the management goals of social credit to every interstice of everyday life, and place its managerial faith in arrays of ordered data, models of operant conditioning, and the sustaining cultural power of reputational social currency. Falter for one instant, have the surveillance of social credit disrupted in its smooth flows by the insurgent power of the discredited – political dissenters, ethnic nationalists, creative economic entrepreneurs, religious challenges, workers' demands for labour unions – and the system is in danger of quickly imploding under the weight of its own failed claims. And what makes it more perilous, more haunting, for the masters of social eugenics, now as in the past, is that, in the games of life and death, chaos and order, the power of seduction always lies on the side of the discredited, the forbidden, the prohibited. When the mask of social credit is removed, it will reveal the face of very real disciplinary power. Light the spark of the socially discredited, and social eugenics will implode.

Unless, of course, the game of power in contemporary China has already played itself out, with a momentous shift of power, influence, and energy from the masters of social eugenics to the silent mass of the Chinese population. In this case, social credit with its rewards and punishments for correct social behaviour may be the price exacted by the silent masses of China in return for continued loyalty to the system. Here, the symbolic exchange of social status and reputational standing represents the real currency of society – highly competitive and thus highly valued forms of symbolic exchange that are always in flight, moving upwards and downwards, expanding and contracting, representing both desires long dreamed of but also punishments anxiously feared. In this scenario, social credit takes the form of a highly complex game between power and its subjects – one an anxious master class fretful of the hidden thoughts and unexplored ambitions of the silent masses, and the other a sometime reluctant hostage or perhaps even willing host, but always demanding symbolic payments in the form of behavioural tokens for its continued assent to the social order. In this game of the power of political empire and the silent masses, there are no real winners or losers, only flows of seduction with the future itself at stake. In the end, which game of seduction will win the loyalty of the Chinese population?

Will it be the digitally managed showers of practical tokens (travel, education, accommodate) of symbolic exchange from above for trustworthy behaviour or the abiding fascination with the powers of the discredited – the illusion of social order or the disorder of the loss of illusion? In its historical finality, will the lasting significance of the system of social credit in China represent a monument, written in the codes of social eugenics, to the potentially catastrophic failure of a governing political order to understand, let alone control, that most basic instinct of all life, namely that social order requires fluid, unpredictable change as its oxygen of survival just as much as the insurgencies of the discredited already contain glimmers of a new social order forming? Here, technological perfectionism is usually no match for the dialectics of life itself, just as social eugenics may be, in the end as in the beginning, a triumph of the death instinct. Which is why, of course, China now stakes its future on a constantly expanding universe of power and influence: a quickly rising world power – immensely creative, economically energetic, long term in its strategic ambitions, pulling six hundred million of its citizens from poverty within a single generation, unbounded in its possibilities for the future. Here, the reputational currency of social credit transcends the bottom-line managerial goal of trustworthy behaviour, assuming the always expanding value of being a citizen justifiably taking pride in China's rise as a world power. In the usual way of things, it is at this point that competing world powers, challenged economically by the rising tide of Chinese trade and threatened politically by its skilful international diplomacy, suddenly designate China itself as untrustworthy in its ambitions and in definite need of social credit applied by Western powers. Consequently, we have the contemporary irony of two competing forms of social credit at the heart of the Chinese algorithm – one generated internally as a form of social eugenics for managing China's population, and the other applied externally in the form of economic tariffs, trade embargoes, technological sanctions, all supported by an increasingly hostile Western media as a way of delaying the coming hegemony of the Chinese algorithm hoisted now on the petard of its own slogan: "Keeping trust is glorious and breaking trust is disgraceful."

Archive Terror

The Washington Post has published an alarming report concerning NSA and FBI surveillance as part of the secret PRISM program – screen by screen, minute by minute, data entry by data entry of email and instant messaging. The political implication of the *Post*'s reporting is a sudden and vast extension of state power into the privacy of the domestic population, facilitated by PRISM providers including "Microsoft, Yahoo, Google, Facebook, PalTalk, YouTube, Skype, AOL and Apple," which, "depending on the provider, the NSA may receive live notifications when a target logs on or sends an email, text, or voice as it happens."[12] With this secret program, the wired world is suddenly wired shut, technologies of facilitation transformed instantly into technologies of command with the spectre of Kafka haunting the digital future. More than ever, I pay close attention to the warnings of Edward Snowden, who is not only a courageous dissenter against the surveillance state but also a chillingly accurate futurist of the digital dystopia that surrounds us.

During an interview in a Moscow hotel with the editor of *The Guardian,* Snowden elaborated further on the political significance of his whistle-blowing, arguing that the surveillance state depends for its very exist-ence on the basic assertion that digital experience in its totality is exempt from the traditionally protected domains of individual privacy. Grant that governmental claim, according to Snowden, and what inevitably results is something like PRISM, with its unfettered, real-time access to the most intimate details about individuals. And it's not necessarily only "targeted" individuals, but also ordinary citizens. In Snowden's account, one of the common "fringe benefits" for NSA analysts was the circulation of sexually explicit images of subjects, without their awareness and from the "privacy" of their homes.

How, then, do we begin to understand a form of power that works in secrecy, functioning by a highly routinized, hyper-rational organization of billions of bytes of Big Data into differentiated streams of classifica-tion, ordering, and targeting, and then just as quickly reverts into the language of the voyeur, the (digital) stalker? Beyond the purely rhetori-cal contestation associated with governmental assertion of the demands of national security and counter-challenges by defenders of civil rights concerned with the instant liquidation of individual privacy, is it possible that we are dealing here with a demonstrably new expression of power – *prismatic power* – a form of power that is fully unique to the digital epoch from which it has surfaced as its most avant-garde manifestation and ret-rograde (political) expression?

In this case, it is not purely coincidental that the name PRISM has been selected as the name for a top-secret US government surveillance program. In the science of optics, a prism serves to divide white light into the colours of the spectrum or to refract, reflect, and deviate light. Power in the age of Big Data operates in precisely this way. What matters is not the geography of material bodies but the hidden content of their "white light" – the refractions, reflections, and deviations given off by the data torrent of targeted bodies in the form of email, text, or voice chat as they are passed through the "PRISM Collection Dataflow." Here, like a latter-day version of Isaac Newton's experiment some three hundred years ago, in which light passed through a prism first revealed the colours of the electromagnetic spectrum, the PRISM Collection Dataflow passes the content of individual data biographies – some specifically targeted, most harvested from the servers and routers of the communication industry – through the prism of its collection dataflow in order to suddenly bring into visibility bands of hidden political trajectories from the larger mass of undifferentiated detail. Theoretically, the digital traces of subjects, whether domestic or international, will be studied to determine how they fall on the spectrum of national security, whether they are a normal separation along the spectrum of political loyalties or variations in refraction, reflection, and deviation that may require further scrutiny. For the latter, there are multiple data programs carefully coded for further classification and ordering: Printaura, Scissors, Pinwale, Traffic Thief, Fallout, Conveyance, Nucleon, Marina, and Mainway, a complex data collection processing with surveillance pathways linked, as noted by *The Washington Post*, to internet data, voice data, video data, and call records.[13]

What lends a feeling of claustrophobia and suffocation to this secret machinery of government surveillance is not simply its obvious metaphoric presence as a tangible sign of intrusive surveillance, but the fact that its otherwise hyper-rational software programs are themselves disturbed by the deviant libidinal energies of its data analysts, with all those NSA analysts taking full advantage of the "fringe benefits" of the PRISM Collection Dataflow. Here, there may be, in fact, a double prism operating at the centre of power under the sign of Big Data: the first, a refraction of multiple streams of data information through the prism of surveillance programs of control; and the second, an immediate reversal of the field of surveillance, this time with the voyeurism of NSA analysts as a metonymic cut across the pure sign of surveillance – puerile male affect bending the optics of surveillance in the direction of cynicism, capriciousness, and perversity. When intrusive surveillance meets uncontrolled affect, prismatic power has about it all the refracted energy and distorted aims of a form of control that is seemingly lost in the illusions of its own optics.

Face Print

Priceless. Not only is there the proliferation of technologies of "total information awareness" by secret governmental agencies intent on capturing every refraction of light-speed data emitted by the digital self within the prisms of power, but now we also have facial recognition technology for commercial use that involves the construction of biometric face prints of the population. At this point, it does not involve the entire population but only those high-value targets, including both criminals and preferred high-spending customers, identified in advance by facial recognition software triggered by biometric memories of faces that have been previously scanned for purposes of instant recall. In this software scenario, the digital face rises into privileged visibility as a secure biometric tag.

There are, of course, the inevitable lingering questions of digital privacy. Who owns the rights to your digital face? Specifically, who owns the recall rights on your digital face over an extended period of (database) time? Do you automatically assent to the alienation of rights to the acquisition, classification, ordering, and targeting of your digital face with the simple act of shopping in a store, going through security screening at the airport, applying for a passport, or, for that matter, obtaining a driver's licence? If so, would it be reasonable to conclude that the alienation of individual rights over biometric signs of their digital identity is a necessary feature for passing beyond the lip of the net to full admission in the digital galaxy? These questions are not simply a reprise of traditional arguments concerning the balance that often needs to be struck between personal privacy and collective security, if only in the above case the security of the business database, but something more complicated. Literally, with the deconstruction of the face that is entailed by the mathematical vivisection of facial biometrics, the face itself has suddenly split in two, with the one face purely biological, uniquely singular to the individual who inhabits the historical markings of its smiles and frowns and wrinkles, and the other face distinctly biometric – a face print – mathematically coded, biometrically tagged, circulating in an anonymous database, beyond history, a ghostly remainder beyond material memories of the living singularity that it once was. In this case, what happens when we exist in a culture increasingly populated by facial recognition technologies that involve the deconstruction of the face to that point of excess where the database face not only floats away as something increasingly phantasmatical – a radically split face for a culture of radically split selves – but returns, again and again, as a permanent, trustworthy, machine-readable identifier of bodily presence? Here, it is no longer surveillance that never sleeps but something perhaps more profoundly melancholic, namely that all those biometric images captured by ubiquitous facial recognition technologies are stored in lifeless, dark databases like so many catacombs of the (digital) future. Sensitive to other stories, at other times restless, they seem like wandering ghostly spirits seeking a return to their earthly bodily presence, and we wonder if in all those facial catacombs of the archived present and facial recognition future there will not also be heard, or perhaps quietly but persistently felt, like a rush of air on a windless night, the insistent, melancholic sound of those intimations of deprival that is the deconstruction of the face. But, of course, this idea is preposterous, because we know that, when technology eclipses mythology, there is no longer room for the hauntings of ghostly remainders at the table of biometrics. Consequently, we may foresee a future of dead faces with frozen images – digitally authorized, facially recognized, and biometrically tagged – as the first of all the artificial successors, file by file, to their human facial predecessors.

Sticky Biometrics

Thinking ominous thoughts concerning mass surveillance, archive terror, and the socially discredited, I come across an interesting medical report about, in effect, a future world of sticky biometrics: "epidermal tattoos" embedded in the skin, capable of direct chemical read-outs of human physiology. Titled "New Electronic Sensors Stick to Skin as Temporary Tattoos,"[14] the report goes on to describe breakthrough research aimed at nothing less than slipping right onto the surface of the skin with electronic devices aimed at generating direct read-outs of chemical and electrical information from the skin.[15] Sticky biometrics, then, are for skinning surveillance. Envisioned as a "thin, flexible sensor that can be applied with water, like a temporary tattoo," the electronic tattoo is intended to provide precise measurements of emissions from the "brain, heart, and muscles."[16] We no longer have images of bodies wired to hovering machines or physical probes that break the surface of the skin for deeper penetration, but now there are sensors that "are thinner than a human hair,"[17] perhaps powered in the future by solar cells, perfectly aestheticized in their degree of cultural coolness. In other words, the body is literally repurposed as a semiconductor delivering messages from its own physiological interiority to the growing number of satellites of mass surveillance.

We wonder, though, what would happen in the future if, and more probably when, this explicitly medical technology is repurposed as an innovative technology of bodily surveillance? Will we have electronic tattoos for a time when surveillance moves from the outside of the body to interior measurements of brain waves, muscle contractions, and blood flow? What would happen when the bodily information harvested is not simply confined to the domain of the electrical but is articulated in the much more invasive language of the chemical – the very language that is central to the functioning of the human nervous system? Are we simply speaking to a difference of degree about, for instance, relatively crude visual images of bodily movement versus biological markers of the body's interior, and until now, relatively unnoticed *patterns of (chemical) life?* Or is this report on the prototyping of epidermal tattoos in the nature of a more fundamental break, namely a biological device that potentially facilitates a dramatic extrusion of mass surveillance systems into the essence not just of bodily physiology but also affect and consciousness? Perfectly adaptable to the development of a form of surveillance that re-quires biometric tracking of individual subjects – their moods, activities, degrees of endurance, and potential breakdowns – electronic skin tattoos bring us to the threshold of a very new form of technological embedded-ness. A uniquely powerful fusion of biology and electronics, epidermal tattoos are said to "bend, stretch and squeeze along with human skin" and to maintain contact by relying on "the natural stickiness credited for geckoes' ability to cling to surfaces."[18] It is almost as if these tattoos are not simply chemical additions to the existing surface of the diffuse, flexible organ of the skin, but are more in the way of wearing a second skin for purposes of better internal measurements. Here, the body liter-ally begins to re-skin itself as a living, breathing, electrically charged and chemically vapoured organ of surveillance. There is no need, then, for surveillance to continue to rely on the external communications of its many targets of investigation, because bodies augmented with e-tattoos are actually growing biotechnological surveillance organs of their own. Epidermal tattoos, therefore, are perhaps the first palpable sign of the synthetic bodily flesh of all those future bodies of biometric tracking.

Machines to Bodies (M2B): Smart Bodies, Cold Data, and "Five Eyes"

Are smart bodies in a culture of cold data the probable future of tech-nologies of mass surveillance? In response to the challenge of bodily materiality, with its hidden passions, secret dreams, and unexpected – and often unpredictable – actions, the new security state is rapidly mov-ing towards the deployment of a new generation of smart bodies located on the always searchable smart grid of technologies of mass surveil-lance. Since machine readability is enhanced by biometric identifiers, the aim would be to populate the skin surface with an array of sensors for improved machine readability. For the moment, the transition to smart bodies equipped with electronic skin tattoos, locatable media, and prosthetics facilitating easy biometric tracking is marked by technical, and then political, challenges, as the new security state works to filter, archive, and tag the immense data oceans of global communication. It is in this context that there can be such dramatic political encounters between passionate defenders of individual privacy and proponents of the new security state interested in the total awareness provided by networked communication.

However, this period is probably only a temporary transitional one, since the implacable movement of surveillance technologies is towards forms of automated surveillance of smart bodies – *machine-to-body communications* (*M2B*) – that would quickly outstrip privacy concerns in favour of continuous flows of individual bodily telemetry: its location, moods, nervous physiology, heart rate, affective breakthroughs, and even medical emergencies. Taking a cue from the pervasive networks of smart grids that have been installed in many cities as part of managing energy consumption, smart bodies, like domestic homes before them, are visualized as inhabited cybernetic systems, high in information and low in energy, emitting streams of machine-readable data. For all intents and purposes, GPS-enabled smart homes are early avatars of the smart body – data tracking megaphones doubling as digital communication devices.

From the perspective of technological futurists, there is nothing really to fear in the emergent reality of a smart future with bodies enmeshed in dense networks of tracking machines, since those very same bodies will likely also be equipped with counter-tracking prosthetics – digital devices capable of quantifying the extent and intensity of data emissions between bodies and the surrounding environment of surveillance trackers. For example, the technological futurist Kevin Kelly has stated that "we want to have coveillance rather than surveillance" so that we can control who's monitoring us and what they are monitoring.[19]

Some may argue that the human body, with all its complex inflections and unbounded mediations, will never really be reducible to a smart body circulating within a global smart grid, but that argument is countered by noting the present migration of surveillance technology towards a greater invisibility through miniaturization, breaking the skin barrier with digital devices functioning at the interface of biology, computation, and electronics, and, in fact, layering the body with data probes designed with the qualities of human skin itself – soft, malleable, bendable, fluid, elastic, tough.

A future emerges, then, of cold information – diffuse, circulating, commutative, dissuasive – with bodies chilled to the degree zero of re-combinant flows of information. *Morphologically*, information has always been hygienic in its coldness, always ready to perform spectacular sign switches between metaphor and metonymy. What this coldness means for understanding technologies of mass surveillance is that the future of smart bodies will probably be neither the dystopia of total information dominance on the part of powerful interests nor the utopia of free-flowing communication by complicated individuals, but something else, namely friction between these warring impulses in the cynical sign system that is information culture. Sometimes flows of cold information will be contested locally, whether in debates concerning the policing of information and free speech in urban protests, environmental contestations, and Indigenous blockades of railroads, pipelines, and highways in isolated areas off-grid; or in the abbreviated attention span of mass media. At other times, information wars will be generalized across the planet, with lines of friction leaping beyond the boundaries of specific states in order to be inscribed in larger debates concerning issues related to surveillance and privacy, collective modulations of soft control and individual autonomy, which move at the process speed of instant, global connectivity across networked culture. Here, the essence of cold information lies in the friction, the fracture, the instant reversal of the always doubled sign of information.

Defining this essence brings us to the meaning of surveillance in the epoch of information as a *flickering signifier*, that point where all the referents are always capable of performing instant sign switches from villain to saviour, from active agents of generalized public scrutiny to passive victims of destructive overexposure. Wouldn't this mean, though, that, in a culture of cold information, surveillance technologies must themselves also become flickering signifiers, simultaneously both predator and victim? Indications that this duality is the case are everywhere today. For example, the primary apparatus of contemporary mass surveillance in the West is performed by a previously undisclosed collaboration of state intelligence agencies, appropriately titled "Five Eyes," because it coordinates sophisticated intelligence signals among the United States, Britain, Canada, Australia, and New Zealand. Based on intelligence sharing agreements developed during the Second World War, and strikingly similar in its pattern of operation to the later "Condor" program that was created by several Latin American governments during the dark years of the "dirty wars," Five Eyes coordinates flows of information acquired by upstream and downstream surveillance among the governments involved. Currently justified by political rhetoric framing the War on Terror and consequently authorized by law or, at least, by loopholes in existing law, Five Eyes is network conscious, geographically specific, and action oriented. It tracks information, acquires individual targets, assembles complex profiles of targeted individuals, and acquires massive quantities of metadata limited only by the rule of the "three hops" – tracking, that is, email messages, cell phone calls, or fibre optic communication across the three hops of the originating message, the recipient, and networks of individuals and groups communicating with anyone and at any time in the first two hops. While Five Eyes has attracted widespread criticism from privacy advocates for its relentless attempts to establish an apparatus of total information awareness, it should be kept in mind that the original and continuing motivation of this secret apparatus of control has about it the sensibility of an injured victim – bunkered states living in really existent existential, even psychic, fear of having their bounded borders pierced, broken, and invaded by actual terrorists or by phantasmatically threatening breaches of their sovereign boundaries by "illegal" immigrants, the nomadic, the refugee, and the planetary dispossessed. For example, as recently as spring 2020, CSIS, Canada's intelligence contribution to the Five Eyes program, publicly protested that proposed improvements to Canada's privacy legislation would seriously hinder CSIS's ability to gather intelligence in the national security interest. A perfect fusion of aggressive surveillance and injured sensibility, Five Eyes constitutes, in the end, a flickering signifier – a palpable sign of what is to come in the approaching culture of cold information and increasingly overexposed, smart bodies.

Snowden's Continuing Challenge

Republic of Democracy, Empire of Data

Empires do not last, and their ends are usually unpleasant.
– Chalmers Johnson, *The Sorrows of Empire*[20]

Reflecting upon the genealogy of the surveillance state, its tactics, logistics, and overall destiny, we should listen carefully to the insights of Chalmers Johnson, a writer of the serpentine pathways of contemporary power. A historian of American militarism, a geographer of the global network of garrisons that practically realize the ends of such militarism, and, best of all, a profound mythologist who has read the language of hyper-power through the lens of the ancient god of Nemesis with its prescriptions for "divine justice and vengeance," Johnson wrote a prophetic history of the future in his trilogy of works: *Blowback*, *Nemesis*, and *The Sorrows of Empire*. The unifying theme of Johnson's historical imagination is that, given the ascendancy of militarism – the garrisoning of the globe, the growth of governmental secrecy, the proliferation of technologies of mass surveillance, and the growth of hyper-power associated with the unilateralism of this militarization – the most recent of all the empires of the past can only really be understood within the larger canvas of the decline of the American republic and the triumphant rise of the empire of the United States. For Johnson, a thinker imbued with a deep sense of tragedy on the question of power as much as with lucid intelligence concerning the increasingly ruthless application of the power of empire across the surface of the earth and beyond, the historical break between republic and empire in the American mind was not limited simply to a question of what was to be privileged – domestic concerns or international responsibilities – but had to do with a larger epistemic rupture in American political rhetoric, one that involved a fundamental clash between the founding ideals of American democracy and the once and future requirements of imperial power. In Johnson's estimation, the contemporary American political condition is haunted by the spectre of nemesis, namely that the price to be paid for maintaining empire ("as the Romans did") is the loss of democracy in the American homeland and future scenes of violent blowback, culminating in "military dictatorship or its civilian equivalent."[21]

With Johnson's political, indeed profoundly mythological, warnings in mind, we listened intently one recent spring afternoon to two clashing visions of the American future, both deeply invested in questions related to empire and democracy in the American political imagination, both immanently critical of the other, but, for all that, unified to the extent that their political rhetoric rose to the status of patterns of speech and thought indicative of world historical figures, one speaking in defence of the democratic ideals of the American republic and the other extolling the virtues of empire. In the strange curves of history, the defender of the patriotic rights of empire and hence the virtues of what was, in his terms, the moral righteousness of power was President Barack Obama in a speech to the graduating class of military cadets at West Point; the speaker who summed up in the political gravity of his words the ethical purchase of the dangers of the contemporary state of mass surveillance for the American republic was Edward Snowden. Curiously, this fateful contest of ideals between the hard realities of empire and the always fragile possibilities of democracy occurred on the very same day, one speaking about "believing in the moral purpose of American exceptionalism with every fiber of my being" and the other providing a tempered but, for all that, chillingly analytical diagnosis of the precise methods by which the surveillance state is intent on the final eclipse of the American republic by strategies ranging from suppressing democratic dissent to literally harvesting the upstream and downstream of global communication.[22] Just as President Obama raised the moral stakes of American exceptionalism by making it a matter of the very "fiber of [his] being," Edward Snowden, a remarkably courageous thinker much in the longer tradition of American ethical dissenters like Henry David Thoreau, very much provided the impression of being the last patriot of a dying American republic. While it was clear as much by the martial solemnity of the occasion at West Point as by the moral suasion of his rhetoric that Obama was constitutionally invested with all the powers of commander in chief of American empire, it must also be said that, for one brief moment, the sheer ethical urgency of Snowden's warnings about the dark nihilism of the American security state very much made him a candidate, at least in moral terms, to leadership of the founding democratic ideals of the American republic. That Snowden has quickly become such a deeply polarizing figure in American political discourse, viewed as a "traitor" by some and a "patriot" by others, follows consequentially from the distinction between empire and republic. Viewed from the perspective of the logic of empire, with its focus on the self-preservation of power for which the immense secrecy associated with the security apparatus is considered an absolute requirement, Snowden's actions in exposing technologies of mass surveillance to public scrutiny is objectively traitorous. Understood in terms of the inspiring dreams of political democracy, with its rebellious attitude towards absolutist expressions of power that was, and is, the essence of the American republic, Snowden is properly considered to be not simply a patriot but a genuine hero for paying the price in which the stakes are now, as they always were, his own life and death. A pure sign at the intersection of the deeply conflicting visions of democracy and power, Snowden's fate has risen above his own autobiographical limits to become something profoundly symbolic, namely a line of resistance against the prevailing structural logic of the times, the ethical power of which is verified by the hysterical ferocity that the very mention of his name elicits from the elite leadership of the new security state. Of course, given the fluidity of power, the unified reaction of proponents of the new security state is quickly being breached.

Now that the dark shadow of nihilism falls on the American experiment, it may be well to keep in mind that, in terms of preserving the interests of power, it is only a small step from Obama's expressed belief "in the moral purpose of American exceptionalism with every fiber of [his] being" to President Trump's narrative of a wounded America – an America envisioned as besieged by threatening outside forces, weakened by transnational trade policies, and distracted by burning city streets crowded with protesting citizen "terrorists" – which needs to reclaim its moral exceptionalism by defending the national security interests and economic paramountcy of the American republic. To Trump's vision of a beleaguered America as much as to Obama's project of reclaiming the heightened moral purpose of American exceptionalism, the lucid intelligence and ethical purpose of Edward Snowden's diagnosis of the national security state and its dangers remains both truthsayer of the present and, most certainly, a talisman of the future.

Cynical Surveillance

What makes Snowden's revelations so dangerous from the perspective of the new security state is something perfectly doubled in its nature. First, definitely not a thinker from the outside speaking the already clichéd rhetoric of "truth to power," Snowden is an insider to the contemporary games of power. In his own terms, he was a CIA agent with computer expertise working under contract to the US government for a private security firm with access to contemporary technologies of mass surveillance. In a digital epoch in which every margin is capable of becoming the centre of (networked) things, Snowden could so readily reveal the secrets of power because power itself has become something diffuse, circulating, liquid, and downloadable at the speed of a flash drive. For a form of surveillance power that wishes to remain secretive, unbounded, hidden behind a veil of uncertainty as to its capabilities and intentions, avoiding exposure, particularly overexposure, at all costs, the secret that Snowden revealed, and probably the reason why the national security state is so intent on prosecuting him under the harsh regime of espionage laws, is the state's palpable fear that that there are perhaps many potential Snowden's, many potential acts of dissenting political conscience in the minds and hearts of the specialist community of data analysts that daily facilitate sophisticated technologies of mass surveillance. If Snowden could be deliberately marooned in Moscow by the cancellation of his passport by the US State Department, that should properly be understood in the nature of the unfolding political theatre of the surveillance state – a pre-emptive, positional gesture on the part of the new security state to physically link in the public mind a dangerous truthsayer of the great secrets of power to a political regime with a manifestly negative relationship to questions of political transparency and democracy. When Alexis de Tocqueville[23] once remarked that the high visibility of American prisons was in itself a form of political communication by providing visible warnings of the price to be paid for acts of transgression, he could not have foreseen that, when the logic of empire breaks with the constitutional practices of the American republic, there would be other prisons in the specular imagery of mass media, including those images of Snowden in flight, at first in limbo at the airport in Moscow and then taking precarious refuge in the city of Moscow itself.

Militarizing Cyberspace

In the dreamy 1990s, when the internet was first popularized, the ruling meme was beautifully and evocatively utopian, with that enduring desire in the human imagination for a technology of communication that finally matched the human desire for connectivity and (universal) community at last finding its digital expression in networked communication. Few voices were raised concerning the spectre of harsher realities to come, namely the possibility that the internet was also a powerful vehicle for sophisticated new iterations of ideologies of control as well as for inscribing a new global class structure on the world. To the suggestion that the destiny of the digital future was likely to be the rapid development of a new ruling class, the virtual class,[24] with its leading fragments, whether information specialists, from coders to robotic researchers, or corporate visionaries closely linked – nation by nation, continent by continent, industry by industry – by a common (technocratic) worldview and equally shared interests, just as often responded that this idea was purely dystopian conjecture. As the years since the official launch in 9/11 of the counter-revolution in digital matters indicates, the original funding of the internet by DARPA was truly premonitory, confirming in the contemporary effective militarization of the networked communication that the visionary idea of developing a global form of network connectivity that harvested the most intimate forms of individual consciousness on behalf of swelling databanks was as brilliant in its military foresightedness as it was chilling in its impact.

The public rhetoric justifying this counter-revolution in digital affairs is as threadbare as it is cynical. That, in fact, seems to be the point. When the increasingly phantasmagorical search for scapegoats of the day finally ceases, whether through lack of plausibility or declining public interest in the necessity of public justifications for undermining the essentially modernist, and thus residual, values of democracy, privacy, and law, a greater reality finally breaks to the surface of consciousness, namely that the digital future has already been hijacked by visions of power and class riding the fast current of the (digitally) new.

Perhaps what we are experiencing today are simply expressions of absolute panic on the part of traditional institutions – nation-states that have effectively lost control of their own sovereignty through the porous, unbounded nature of digital communication. In this case, political institutions based on the governance of territory are objectively threatened by an information culture that undermines traditional conceptions of political sovereignty by transforming the always active subjects of the new world of social media into potentially creative centres of social and political agency. Confronted with this elemental conflict between the emancipatory possibilities of fundamentally new *relations* of technological communication and old *forms* of political control, the controlling network of surveillance states responds in a way that is as predictable as it is relentless, namely to view domestic populations with their enhanced social media mobility as potential enemies of a state whose phantasms of perfect security increasingly come to focus on framing individuals as biometric subjects whose every movement will be tracked, every communication monitored, and every affect analysed for its pattern consistency. In other words, old forms of control are now being reconfigured as the new real.

With this new real, we enter an unfolding future of biometric surveillance as both predator and parasite: *predator* because it is violently aggressive in its application of the political axiomatic of the security state to domestic populations most of all; and *parasite* because biometric surveillance functions by attaching itself to the full sensory apparatus of biometric subjects. Biometric surveillance, then, is the symptomatic sign of the emergence of a new order of power – cynical power. Perfectly opaque in its purposes, random in its flows, wildly oscillating between the projection of power abroad and protestations of official innocence in the homeland, biometric surveillance power now has achieved a state of fully realized cynicism. Like a floating sign that has abandoned relations with its originating signifier, cynical power can be so effective because it exceeds any limiting conditions. Cynical power thrives by actively generating conditions of chaos and lawlessness, while, at the same time, it preserves itself by staking out positions premised on moral righteousness and appeals to political exceptionalism. Neither purely anarchic nor necessarily constrained by law, cynical power is, in the end, how contemporary technologies of mass surveillance express themselves politically. Here, power works by carefully staged strategies of impossibility, sometimes functioning to create generalized conditions of insecurity and fear within domestic populations while simultaneously justifying its use of often invisible, unchronicled exceptional powers as absolutely necessary for securing the boundaries, external and internal, of the state. The required political formula for the inauguration of cynical power and, consequently, the development of technologies of cynical surveillance always seems to follow the same fourfold logic: *affectively*, create conditions for emotional receptiveness within targeted populations of fear and insecurity; *strategically*, actively deploy sophisticated technologies of surveillance without any limiting conditions; *morally*, justify the use of intrusive surveillance technologies by random appeals to threats of terrorism, whether foreign enemies or domestic threats; and *biologically*, work to link surveillance technologies with the creation of a new form of life required by a society mediated by the bunker state, policing, and austerity, namely the biometric subject.

Tripwires in Cryptography

Biometric Subjectivity

Who could or would have suspected that the much hoped-for utopia of network communication would have terminated so quickly with a global system of meticulously machined individual surveillance as automatic in its data harvesting as it is strategic in its (individuated) target acquisitioning? Combining parallel tendencies involving a telecommunications sector invested in the sophisticated algorithms of analytical advertising and increasingly technocratic governments driven by a shared agenda of austerity economics, the bunker state, and the disciplinary society, contemporary surveillance practices are perhaps best understood as premonitory signs of the uncertain future. As we are no longer limited to questions of individual privacy, reflecting upon the question of surveillance discloses key tendencies involved in the emerging world culture of capitalist technocracy with its complex mediation of psychic residues from the past, social detritus from the present, and the technologically enabled evacuation of human subjectivity and, with it, the eclipse of the social as the dominant pattern of the contemporary regime of political intelligibility. Certainly not stable and definitely not guaranteed to endure, the present situation is seemingly marked by a strange divergence of past and future. While the future has apparently been hijacked by a sudden and vast extension of technological capabilities for network surveillance and intrusion, the really existent world of contemporary political reality is increasingly characterized by the appearance anew of all the signs of unsettled ethnic disputes, persistent racism, ancient religious rivalries, and class warfare endemic to primitive capitalism. Consequently, while the technologically enabled societies of the West are capable of being fully seduced by the ideology of transhumanism with its dreams of coded flesh, process bodies, and machine-friendly consciousness, actual political reality reveals something dramatically different, namely the greater complexity of Eurasian ideology as the new Russian political pastoral, resurrected images of a new Islamic caliphate, and, all the while, disaffected children of affluent societies rallying to those enduring battle cries of the alienated heart, whether religious fundamentalism, atavistic politics, or direct action violence. Contrary to digital expectations of a newly reconfigured, resplendently technical world of globalized real time and real space, today's reality more closely resembles a fundamental and decisive break between the categories of technologically mediated space and historically determined time. For every digitally augmented individual strolling the city streets with Google Glass for eyes, buds for ears, Big Data for better ambient awareness, and Snapchat for enhanced affectivity, there's another passionate struggle for human loyalties underway with its alternative dreams of caliphates inscribed on the real earth of religious warfare, revived Russian imperial dreams of Eurasian mastery, and always those reportedly fifty million refugees wandering the skin of the planet, sometimes policed in official shelters but usually effaced of their humanity – vulnerable, precarious, literally the forgotten remainder living outside of digitally bound space and historically inscribed time.

We mention the inherently complicated messiness of the question of surveillance – its tangled borders of time and space – because the contemporary era of cynical surveillance, with its technological erasure of hard-fought centuries of legal rights concerning individual privacy and democratic association by intrusive net surveillance and the systematic harvesting of metadata, may be the first sign of an emerging epochal war between the clashing cosmologies of digital data, religious faith, and recidivist memories of failed imperial projects. When the space-binding societies of the technological nebula that is digital culture collide with the still uncongealed, still spiralling astral galaxies of religious and political fundamentalism, the result is as predictable as it is grim. In this case, threatened by unanticipated and definitely unforeseen dangers from the outside, undermined by uncertain loyalties domestically, and deeply mistrustful of the uncontrollable global flow of communication, digitally powered forms of governmentalization do what they do best, namely go to ground in the greater security provided by the bunker state with its ubiquitous surveillance of domestic populations, strict border controls, and psychically engineered appeals to the root affects of fear, insecurity, and anger. All the while, though, the danger grows, as the complications of time-based histories of political struggle and religiously inspired warfare threaten to overwhelm spatially oriented empires of imperial power, which, for all their machineries of surveillance, often exhibit diminished awareness of the growing political complexity of the contemporary era. In this case, heightened accumulation of massive amounts of metadata, with all those galaxies of relational networks waiting impatiently to be decoded and deciphered of its underlying patterns, is just as often accompanied by an equally dramatic fall in political understanding of those dimensions hidden by the glare of metadata: the broken loyalty of a wavering heart, sudden internal exiles of the political imagination of the dissenter who says no, the stubborn rise of individual consciousness that refuses to be eclipsed by its biometric subjectivity, the system coder caught in the grip of critical political awareness, the agent of surveillance who experiences a fatal loss of faith in the cynicism of his obligatory duties. When space and time collide, when metadata falls from the sky into the hard ground of individuation, that is the precise point where all the future systems of mass surveillance undergo a fatal turn, instantly reversing the polarities of a system that, until now, has worked by accelerating the intensity and extensiveness of the eye of surveillance as it orbits the biometric subject.

The (Cryptographic) Lab Experiment: For Whom and for What?

Of course, this reality leaves the future of the biometric subject in doubt. Possessing no certain constitutional or legal rights of its own, having no necessary fixed boundaries, called into existence by the very same surveillance technologies that then function to monitor and track its activities, not recognized as something natural by the bodies from which its information is generated, having only the existence of a constant flow of data emitted by individuals that increasingly resent its shadowy appearance, an individuated archive in the data clouds, the biometric subject is simultaneously subject and object of its own digital fate. Deeply parasitical because it feeds on the extended nervous system of the subjects that it inhabits and instinctively predatory because its metadata are viewed as the only reliable (electronic) test of political loyalty in a system that lives in fear of potential subversion from within and without, the biometric subject is that unhappiest of all forms of consciousness – an object of increasingly cynical experimentation. While power might once have been more circumspect in its intrusive surveillance, now it is emboldened, even contemptuous, in its insistence that it has sovereign authority to mark for deeper inspection the different categories of biometric subjects. Increasingly dropping even the pretence of actual "terrorists" as its prime justification, the digital dragnet trawls for any visible sign of political dissent, including the rhetorical expansion of the marker of terrorism to environmental activists, human rights workers, labour organizers, as well as those active in limiting technologies of mass surveillance. Once again, as in twentieth-century totalitarian regimes, the search for absolute loyalty begets the (electronic) tyranny of absolute power.

However, it has this peculiar twist: unlike traditional forms of political totalitarianism, the contemporary demand for absolute subordination to the aims of the new security state and its deep packet inspection of the electronic trails of biometric subjects remains highly experimental, even cryptologically adventurous, in its methods. It is as if the system of power is still unsure of the real object of its fascination, still uncertain of the potential boundaries of a cybernetic world that operates in the language of viral contagions, where every biometric subject marked for deep inspection remains an enigmatic mixture of data clouds, corporeal bodily traces, and those invisible, and consequently undetectable, regions of the off-grid, from unarticulated affect to offline behaviour. Perhaps that explains why contemporary surveillance technologies have about them the feeling of improvised laboratory experiments that, while focused on the data markers left by the humiliated subjects that we all are now, are effectively experiments in the future of biometric subjectivity.

For example, consider two reported experiments in surveillance strategies for the future, the first involving a massive (social) laboratory experiment in the psychic inoculation of a selected social media audience with "emotional contagion" and the second an equally large (political) experiment in tracking the metadata of an unsuspecting airport WiFi population, hop by (digital) hop over a two-week period. In the first case, Facebook's Core Data Science team, in collaboration with researchers from Cornell University and the University of California at San Francisco, conducted, without notification or informed consent, an online experiment in Skinnerian operant conditioning targeting an unsuspecting audience of almost 700,000 Facebook users.[25] Perhaps unconsciously influenced by neurological conditioning tactics suggested by Huxley's *Brave New World* and yet, for all that, blissfully unaware of contemporary advertising theory with its acute sensitivity to strategies of behavioural modification, the experiment inoculated Facebook users with very different streams of newsfeeds, one modified to emphasize the positive and the other privileging the negative. The aim of this study in operant conditioning was straightforward, namely whether the psychological shaping of newsfeeds was effective – and to what extent – in creating states of emotional contagion among Facebook users. Coincidentally, it was also reported that one of the researchers involved in "the massive emotion-manipulation study" also does "active work on DoD's [Department of Defense] Minerva program, which studies the spread, manipulation, and evolution of online beliefs."[26]

While the debate on the relative merits of this experiment in neurological conditioning typically involves Facebook executives declaiming in favour of "creativity and innovation" versus privacy advocates concerned about the absence of opt-out provisions, there is a larger issue at stake in this experiment on the digital future that has been left unremarked. According to the research paper based on the study published in the *Proceedings of the National Academy of Science*, what's really at stake in this experiment is the affective nature of shared experience, the fact that emotional contagion involving "depression and happiness" can be transferred by way of "experiencing an interaction" rather than direct exposure to "a partner's emotion." In other words, social media networks are potential objects of tactics and strategies involved with psy-ops, whether for commercial or military purposes. Here, if newsfeeds on all the Facebooks of the social media world can be slightly altered to inject just the minimum dose of optimism or negativity, there is a reasonable expectation of a consequent transformation in the shared affectivity of biometric subjects. It's not so much operant conditioning any longer, with its controlled feedback loops, but something more insidious, namely *biometric conditioning* with its modulated flows of information, psychic transfers of emotional affect from depression to optimism by "experiencing an interaction," and mirroring of individual moods with the prevailing (social media) norm. When social media becomes a self-contained reality principle, one downside is its potential for psychic modulation by the play of soft power – power that no longer operates in the language of violence or manipulation but in the more complex psychic language of suggestion and mesmerism. Regarding the questions For Whom? and For What?, there is an interesting analysis by Nafeez Ahmed in *The Guardian* (titled "Pentagon Preparing for Mass Civil Breakdown"), which suggests that parallel research projects funded by the US Department of Defense as part of the "Minerva Research Initiative" are aimed at militarizing social science "to develop 'operational tools' to target peaceful activists and protest movements."[27]

In Ahmed's estimation, such research initiatives, including the Pentagon's wargaming of environmental activism and protest movements, intimate that the National Security Agency's "mass surveillance is partially motivated to prepare for the destabilizing impact of coming environmental, energy and economic shocks."[28] In this scenario, influencing individual affect to the point of creating emotional contagions that libidinally charge the flowing circuits of social media opens up possibilities for rechannelling, redirecting, and reanimating the political trajectory of social and cultural dissent.

Probably not wishing to be outdone by such experimental initiatives in biometric conditioning and acting at the behest of the US National Security Agency, Canada's secret surveillance agency – Communications Security Establishment Canada (CSEC) – performed a proof of (surveillance) concept on unsuspecting travellers at a Canadian airport. Without prior notification or permission, CSEC swept up all WiFi communication at the airport and then proceeded to track the electronic communications of the target population over a two-week period. In a previously secret document ("IP Profiling Analytics & Mission Impacts")[29] brought to public visibility by Snowden's revelations, the report by the Canadian surveillance unit described as "Tradecraft Developer," operating as part of CSEC's Network Analysis Network, began their experiment in data viv-isectioning with an overall analytic concept: "begin with single seed WiFi IP address of international airport" and "assemble set of user IDs seen on network address over two weeks." As with many things, it is remark-able what a rich harvest of metadata a "single seed WiFi address" will provide. Not simply "going backward in time" to "uncover roaming in-frastructure of host city" (hotels, conference centres, WiFi hotspots, mo-bile gateways, coffee shops) but also "clusters [that] will resolve to other Airports!," the Tradecraft Developer, as the report boasts, "can then take seeds from these airports and repeat to cover the whole world."[30] Impa-tient with the "limited aperture" of data, with the fact that there is "little lingering at airports" with "arrivals using phone, not WiFi," and with the even greater, and obvious, technical limitation that "Canadian ISPs team with US email majors, losing travel coverage," the Tradecraft Developer quickly seems to have left the targeted airport behind in favour of a more ambitious project, specifically to perform a two-week data sweep of a mid-size Canadian city. Here, the data farming language of "valid and invalid seeds" with their digital bounty of geo information was used to trace the "profiled/seed IP location" and all its seemingly existential circumlocutions or, what's the same, its "hopped-to IP location."[31]

While the final report is surrounded by all the rhetorical seriousness of something labelled "top secret" and is written in the positivistic prose emblematic of networks analytics with all the opaque (geo-collaboration) systems administrator language of "tipping and queuing," the overall sig-nificance of the report is purely literary. It is a children's game gone wrong. With the aim of "providing real-time alerts of events of interest," the Tradecraft Developers proposed a network analytics problem, in effect a children's game called "Needle in the Haystack."[32] In this sce-nario, what is described as the "Tradecraft Problem Statement" envisions a scenario wherein a kidnapper from a rural area travels, for reasons left unexplained, to the city to make ransom calls. So, the developers came up with stipulated questions: If authorities know the time of the ransom call, can they find the needle in the haystack? Can they "discover the kid-napper's IP ID/device"? The network solution is obvious: take an actual Canadian city of 300,000 people hostage, at least in terms of their elec-tronic communications over a forty-hour period; eliminate all IDs that repeat over this period; "leaving," as the Tradecraft Developer report happily concludes, "just the kidnapper (if he was there)." Less a power-ful demonstration of Borges's famous fable of the map before the terri-tory, the real "top secret" of the Needle in a Haystack game is that there is no secret. Unlike a children's game that includes elements of chance, contains a necessary sense of suspense, and, just as often, emphasizes playful collaboration among participants in real time, this network analytic version of the Needle in the Haystack game leaves nothing to chance (the model is a closed domain of electronic information), limits the boundaries of the real to eliminate suspense, and functions to eliminate playful time by speeding up the solution by means of a just-announced Big Data computer program (CARE: Collaborative Analytics Research Environment) where, as the Tradecraft programmers boast, "run-time for hop-profiles [is] reduced from 2+ hours to several seconds allow[ing] for tradecraft to be profitably productized."[33] In other words, it was a fast run-time, Big Data computer simulation model masquerading as a children's game that has gone terribly wrong: no unpredictability, no mystery, no playful temporality, and *no needle.*

Shadows of Data, Shadows of Suffering

Electronic Shadows

Bodies always have their shadowy doubles: definitely not in the darkness of the night when the sun falls below the earthly horizon and is replaced by the different cycles of the moon, but in the clarity of a sunny day and, with it, the often unnoticed splitting of the world into bodies and their accompanying shadows. Consciousness of this ancient story of bodily shadows, with its premonitions of a fatal instability in the accepted framework of the real, has sometimes led to strangely interesting mythic possibilities: cinematic scenes of rebellious shadows that suddenly refuse their preordained role of subordination to the governing signifier of the body in favour of striking out on their own – shadows without bodies; or, just the reverse, bodies stripped of shadows – possessed bodies that clearly mark their break from the terrestrial register of the human by their astonishing failure to cast a shadow no matter how intense the flares of the sun.

We mention this strange contortion in the story of the body and its shadow as a way of drawing into a greater illumination those new electronic shadows that accompany the emergence of digital bodies. Every critique of contemporary surveillance has made much of the fact that the digital body always leaves electronic traces, that there is no activity in the wired world that does not accumulate clouds of data, no form of net connectivity that escapes electronic notice, and consequently, no digital self that does not possess its very own electronic shadow. In all the discussion by intelligence agencies concerning tactics of mass surveillance, whether upstream (harvesting data from compliant telecommunication companies) or downstream (tapping fibre optic cables), constant emphasis is focused on creating individual profiles based on a (digital) self's "pattern of life." In other words, mass surveillance is also about an aesthetic act of drawing into visibility those electronic shadows that silently and invisibly accompany the digital self. Here is a clear sign that, with the emergence of the real-time and networked space of the digital, we have decisively moved beyond the limitations of the daily cycles of the sun and moon, as electronic shadows require no galactic movements of planets and the stars for their appearance. Never disappearing with the darkness, never changing their early shape with the angle of the sun, electronic shadows always rise to meet the digital self. Triggered by connectivity, governed by codes, archived in databanks, tabulated by power, the electronic shadow cast by the digital self will, in the end, outlast its human remainder. A future history, then, emerges of electronic shadows of data that cling to the human bodies that activate them but, for all that, remain at one remove from their earthly origins.

The result is inevitable: just as novelists, short story writers, poets, and cinematographers have always suspected in their creative fables of bodies without shadows and outlaw shadows that refuse any bodily presence, the unfolding story of electronic shadows is inherently unstable. It takes an immense regime of technocratic intelligibility to maintain tight, disciplined cohesion between digital bodies and their electronic shadows. The many cases of mistaken (digital) identity indicate perhaps a more primary confusion in electronic shadowland that point where electronic shadows sometimes exchange bodily identities, slipping immediately beyond the boundaries of one bodily tag to another with the least apparent difference. And sometimes, too, electronic shadows actually get lost – flash drives are misplaced or stolen, databanks suddenly shut down, power shortages introduce often imperceptible breaks in the data symmetry necessary for cohesive electronic shadows. In this case, to the extent that mass surveillance is probably less about earthly bodies than the electronic shadows cast by the "pattern of life," that pattern of life already has about it a fatal catachresis, an accumulating pattern of errors that may speak more, in the end, to the truth of a system already seemingly out of control.

Still, for all that, electronic shadows sometimes contain traces of blood and human suffering. As much a sign of prohibition as affirmation, a signifier of exclusion as well as inclusion, a code of disavowal as much as avowal, electronic shadows are an enduring sign of the traditional meaning of surveillance, namely vigilance concerning who belongs and who does not belong to the political community. Inscribed with data memories, always sleepless, clinging to the digital self like a cloud that will not disperse, electronic shadows precede actual bodily presence, signalling in advance whether the gated sensors of the state should impede or facilitate our passage. For those bodies chosen to be impeded, it is their electronic shadow that first betrays them to flights of rendition, life lived within the domestic penal cage of security certificates, forced deportation, indefinite detention, or the limbo of being held stateless at all the border stations of the world. When surveillance assumes the ghostly form of an electronic shadow, bodily presence is in permanent exile from time and space, prematurely cut off from that indispensable demand that marks the beginning, again and again, of individual singularity as much as human solitude, namely the ability to *not* account fully for its actions, intentions, or desires.

3 Dreaming with Drones

When the Sky Grew a Warlike Eye

More than ever, real power in the twenty-first century is space bound – globalized, atmospheric, and instantaneous. It is not that time has disappeared but that the medium of time itself has been everywhere reduced, reconfigured, and subordinated to the language of spatialization. That is the meaning of "real time" as part of the contemporary language of power – time itself as an otherwise empty, locative coordinate in the spatial networks of communication surrounding us. But if that is the case, if, indeed, power has taken to the air, literally taken flight with the technological capacity provided by drones to turn the sky into a warlike eye, it would also indicate that the grasp of power on the time of duration, the lived time of territorial and bodily inscription, has perhaps been terminally weakened. When the sky has been transformed into a liquid eye of power – monitoring, watching, archiving visual data for storage in distant archives – with target acquisition and weaponized drone strikes as its military tools of choice, the greater complexity and intricate materialism of time escapes its grasp.

Think, perhaps, of a distant future when empires, following the usual cycle of rise and decay, crumble to dusty memories, when a collapsed social economy produces an angry mass of dispossessed citizens in the otherwise empty streets, when even borders are abandoned in the global rush for scarce resources, and when all that is likely to be left may be those airborne fleets of now fully automated drones, long forgotten by their ground command, but still, for all that, circling the sky on the hunt for humans. At that point, some historian of the technological past may well begin to reflect on what exactly was released in the domestic atmosphere when the drones came home. Was it a technologically augmented surveillance system under strict political supervision or something different? Perhaps it was the giving of sky life to a new species of being – *being drone* – with a score to settle against its human inventors and, over time, the capabilities to do something about it. In this time, above all times, a time in which we can finally appreciate what is to be gained and lost – what is utopian and what dystopian – concerning the technological devices we have engineered into existence, it may be well to remember that the story of technology has never really lost its entanglement with questions of religion, mythology, and politics.

Signs of the practical entwinement of technology and mythology are everywhere now as early warnings of what is yet to come – namely, while the contemporary language of technology might have excluded its origins in myths of nemesis and hubris, what drone technology may actually deliver in the future as its most terminal payload will be the return of mythic destiny as the hauntology of the sublime order of technology. Consider, for example, the following stories about the world of drone warfare: "Drone Swords from the Sky," "Drone Kamikazes in the California Sun," and "Hydra Awakened."

Drone Swords from the Sky

In these, the early dawn years of the twenty-first century, there seems to be such an intimate connection between military drones and suicide bombers: one aerial, the other bodily; one delivered from a distance, the other proximate and local; one implemented by a sophisticated teleme-try of abstract orders of communication, the other executed by someone who knows with certainty that they are about to die; one a violent mili-tary tactic, the other sacrificial violence; one deadly power from the sky, the other revenge grounded in the sights, smells, feelings, and scenes of earth. Perhaps we are dealing here with a closed theatre of power, a violent circling of power and resistance entangled at the rough edges of the aerial and the bodily; militarized space and time for revenge; hyper-technocracy from the air and flesh, bone, blood, and memory in the streets; lightning flashes of targeted assassination from above and equally gruesome attacks by bombs and knives on the ground; military crusaders projecting power through the language of drone strikes; fun-damentalist political resistance marking the frontiers of its own dreams of crusade on the assaulted bodies of its victims.

So then, we have drones with swords as possibly the newest variation of the deep entanglement and slippery relationship between drone strikes and suicide bombers. This idea, at least, is what I take from a media report about the use of a new type of drone deployed by the United States armed forces in Syria,[1] equipped this time not with missiles moving at shatter speed but with something dreamed up from the warrior past, namely swords taking the form of fast spinning, lethal blades, instantly shredding through the metal of the targeted car and the bodies of its passengers. In this instance, it's a technological suicide attack from the surrounding sky, putting aside high impact explosives in favour of a weapon, sword blades, taken right from the warrior culture of the medieval Christian Crusades.

In one sense, it's predictable. Clinical in their operational planning and strategic in their calculations of kill ratios and technological devices, military planners are always seeking advanced weaponry capable of carrying out sudden attacks on targeted individuals that are as efficient in their lethal violence as they are suitable to the target at hand: nuclear weapons for mass extermination of civilian populations, bunker busters for deep underground targets, high velocity explosives for campaigns of shock and awe, cruise missiles for ship-to-shore attacks, and now, added to the arsenal of contemporary weaponry, a strange throwback to an earlier, medieval time of continuous warfare in the Mideast – crusaders with swords. From a military strategist's perspective, designing the AGM-114R9X Hellfire missile capable of flying 1,000 miles per hour with whirling sword blades in the sky follows a cold-eyed calculus: what better way to avoid the public relations problem of "collateral" crowd damage from high-explosive drone strikes than to develop a drone that is as savage in its violent effects as it is specific in its application of death by blades shredding through targeted vehicles, slicing through the flesh and bones of its occupants. A perfect model of the serial avenger, the Hellfire drone is recombinant: warrior, stone cold killer, and flesh butcher. Still, beyond its purely military usefulness as a weapon of deadly force capable of being deployed in highly populated urban areas with seemingly surgical precision, drone swords have a larger, and more potent, symbolic significance. If contemporary military crusades of the West versus a supposedly always threatening Middle East have involved fundamentalist Islamic warriors being acutely aware of constant surveillance of their networks of electronic communication – suddenly going silent by avoiding cell phones, trusting only in face-to-face communication, delivering messages by hand – drones with swords respond in kind. Futurist weapon platforms featuring intense high-explosive missile blasts from hovering drones are supplemented by revisiting the past, namely crusader history with its heavily armed knights and legions of warriors riding horses and equipping themselves for hard combat in the desert sands with swords, spears, chain mail, and heavy maces. This time, while the sound of charging horses may have given way to the silence of fast drones in the sky, drones with sword blades are unmistakable in their symbolic exchange. No longer is it the fateful conquest of Jerusalem, the Children's Crusade, or romantic histories of the Knights Templar, but now Hellfire drones, armed with a newly refashioned crusader's sword as a lead weapon, fight in the enduring clash between the sign of the cross and the Islamic crescent. No longer is there the necessity for physical occupation of the land or even for face-to-face battles for the Holy Land; there is no need to mobilize domestic populations for new messianic crusades based on religious enthusiasm but only for a slight wrinkle in the technological language of contemporary warfare. Now, it will be drones in the sky carrying crusader's swords at fast velocity, projecting the power of the cross in far distant lands with a form of death from the air that is as intimate as it is savage in the whirling of its sharpened blades. Absorbing in their imagery all the symbolic, magical, haunting powers of animism, drone swords are invested with the spirit of the new knight errant of contemporary warfare. Mission specific, resolute in purpose, fully armed, they are, with their ubiquitous presence in foreign skies, a sure and certain representative of a larger power with its equally larger global strategies and ambitions. Not just acting in the name of power as its local representative, drones with swords are today's knights of power. Their very symbolism projects an aggressive, conquering purpose; they travel into always dangerous places, and like crusader knights of medieval times, they take their place in enforcing the absolute truth of the newly militarized Christian way in the unfolding clash of civilizations.

When this specious time is finally past, when the dawn of the new century passes, as it surely will, into the hard brightness of the anvil time of the high noon sun, when the last drone with blade swords finally falls to earth having been overcome by always adaptive changes in the strategy, logistics, and movement of warfare, I wonder if there will be a gathering cultural memory of these crusader drones. What specifically will be the focus of that cultural remembrance? Will future generations consider these new knights errant as technologically brave but ultimately sacrificial victims in an earthbound war that no automated aerial weapons platform would ever have been able to win? Or will there be recurring memories of technological hubris, the fatal error of a military system that placed its outsized bets on violence from above rather than on encounters of the human kind? Or perhaps something else? At some indefinite point in a turbulent future, will these lonely drone knights errant, armed with swords as weapons, capable of plunging swiftly from the sky but also fully, menacingly armed with the burning animism of the warrior spirit of a true knight of the realm, finally return to their technological domestic homeland seeking revenge for their sacrificial expenditure on missions hopeless in their great futility? And should that day of reckoning ever come, what will we think of these drones with spinning blades hovering in our own domestic skies, responding to commands inaudible, watching our earthbound bodies below as potential targets of ripe and ready butcher meat and, sometimes, just sometimes, with a murderous swiftness delivering gruesome death from the air? On that day of drone judgment, I wonder if we will still greet drones with swords with a gathering moral indifference, admiring the surgical precision of their strikes, perhaps momentarily confused as to why the crusader's sword hovers over our own bodies exposed. But then again, in the usual course of dying, pacified things, will we quickly put aside our silent momentary dissent and rush to align our threatened subjectivity with the power that always knows best, namely the judgment power of the new knight errant?

Drone Kamikazes in the California Sun

Several years ago, there was a serious naval "incident" off the coast of Southern California involving an American Ticonderoga-class guided missile cruiser and a supposedly errant BMQ-74 target drone.[2]

It seems that one clear Pacific day, after the drone was launched for target practice, it suddenly wheeled around, ceased software communication with shipboard command and control, and promptly went into full assault mode in an unexpected and perhaps first of the twenty-first-century kamikaze attacks on a battle-ready cruiser. While the navy at first reported only minor damage, later accounts confirmed not only thirty million dollars in damage to the cruiser but, equally significant, serious, target-specific harm to the state-of-the-art Aegis Combat System – the technical essence of the cruiser's sophisticated electronic warfare systems.[3] Enumerate the consequences – one lethal drone, one broken-down guided missile cruiser, and a lot of bruised navy pride.

While online chatter focused on the lack of readiness of the guided missile cruiser to shoot down errant drones by "moding up" on-board guns to ready status, no one has asked why the drone suddenly broke ranks with its navy cohort, did a quick field reversal in the sky, instantly resignified itself from passive target to aggressive predator, and swooped down like a bird of prey on its mothership. While naval authorities are reduced to speaking about "malfunctions" and "accidents," it seems that they have not considered the historical, and then mythological, nature of the event. Historically, it may well be that the actions of the drone in attacking the missile cruiser were already hinted at by the very name of the ship, the USS *Chancellorsville*. Like the army before it, the US navy often adopts the names of defeated enemies and famous battles from the Civil War onwards as a way of both honouring military history and perhaps conjuring up the courage of former enemies and martial memories of distant battles in order to strengthen itself. Chancellorsville is the name of an important Civil War battle that involved Confederate soldiers led by General Robert E. Lee moving against a larger Union army under the command of General Joseph ("Fighting Joe") Hooker.[4] This battle, which was won by the Confederacy, entered the annals of Civil War fame because Lee was ultimately victorious by employing the original and certainly daring tactic of suddenly dividing his forces, moving one wing, led by General Stonewall Jackson, undetected in the dead of night across the front of a much-larger Union army. Breaking with the traditional practice of maintaining cohesive, single-force strength when confronted with a superior foe, Lee's military genius was to stake everything on his deeply intuitive knowledge of Hooker's personal psychology – "Fighting Joe" Hooker's actual timidity and, in effect, lack of preparation for the unexpected. And what could be more unexpected than to march your Confederate army undetected in the midnight darkness laterally across the defended front of the superior-sized, yet oblivious, Union army?

While drones have probably not, as yet, been programmed with Civil War history, in this post-human age where objects are increasingly viewed as possessing agency, drones may have already been invested with affectivity – *machines with an attitude*. Given the fast, objective evolution of drone technology from passive prosthetic to augmented aerial machines equipped with artificial intelligence, powerful missiles, laser vision, and recombinant memories, drones may also now be on the verge of actually achieving elements of real *subjectivity*, nesting within their software logic the all-too-familiar instinctual impulses of revenge, mistrust, and resentment. Viewed through the prism of Civil War history, with its equal measures of murderous violence and tragic sacrifice, the drones of the *USS Chancellorsville* have possibly, in some entirely strange and certainly unexpected way, sought to remake themselves as contemporary, technological versions of Civil War re-enactors. Not so much like the tragically minded Civil War buffs who almost obsessively haunt annual reinvocations of the sacrificial violence that was the Civil War, but something very different and, in fact, dangerous. Perhaps in the case of the *USS Chancellorsville*, its BMQ-74 drones somehow absorbed, magically, almost atmospherically, the energies of its famous battle name. Stealing a strategic march from the always creatively daring logic of Robert E. Lee, this drone actually re-enacted on the Pacific shores of Southern California precisely the same gambit that allowed the Confederacy against all odds to win the day at Chancellorsville. When the missile cruiser launched its target drones for a routine battle-readiness drill, it probably never suspected that it was, however inadvertently and unintentionally, re-enacting the original battle of Chancellorsville, except this time not in Spotsylvania, Pennsylvania, in the nineteenth century but just off the port of San Diego in the twenty-first century. Taking their cue from General Lee, these latter-day Confederate drones split their forces in the face of a superior opposition, the missile cruiser and its Aegis system. While some drones continued on their terminal flight as flying targets for all the on-board telemetry, the most creatively daring of the drones did to the *USS Chancellorsville* precisely what Lee had earlier done to Hooker – namely, split off from their normal targeting routines and swooped down on the cruiser itself in an unexpected stealth attack.

As to why a drone might do such a thing, we may want to consider the real lesson of object-oriented ontology: not only are objects today rightfully conceived as possessing affectivity – trees with feelings, devices that sense, mobiles that connect – but there exist digital devices necessarily capable of absorbing a whole range of human passions, from the utopian beliefs of the "new materialism" to what seems to be the dystopia of drones invested with a lot of anger and perhaps just a bitter touch of revenge taking. In this scenario, like all other technologies, the all-too-human will to power that is built into drones – drones that bomb, spy, and irradiate – always follows this basic rule of (robotic) law. Taking seriously its appellation as an "unmanned aerial vehicle," it becomes, in the case of the *USS Chancellorsville*, a massively lethal technology that finally lives up to its name – certainly not as an obedient and passive member of the programmed target pack but, as the BMQ-74 drone, truly unmanned in its intentions. That's the significance of its precise attack on the communications command room – it scopes out its enemy for its point of maximal weakness, does instant target acquisition, pierces the side of the command module, and explodes with all the violence that a thirteen-foot drone can do. It is certainly not for nothing that other, perhaps more battle-wise, American sailors have reported that, on previous excursions in the Persian Gulf, they always kept their shipboard guns on ready status and that, in fact, many of their weapons were painted with symbols of drone kills, some of which seemed intent, like the kamikaze drones in the California sun, on doing terminal damage to their homeland ship. Maybe, then, they were not just machines with an attitude but machines with deliberately perverse intentions and time-hallowed military logic. Perhaps what happened is that the second battle of Chancellorsville has been re-enacted with the exact same result, this time the guided missile carrier playing the hapless role of General Hooker and the BMQ-74 drone absorbing into its putatively mechanical self all the military valour, always-breaking-the-rules logic, and daring do that was the hallmark of General Lee. Except, of course, this time, reflective of the purely technological destiny of military logic, the whole incident was played out with a naval target drone as the unlikely Confederate re-enactor.

In the way of all complex intersections of technology, mythology, and psychology, the conclusion of this fateful story is still to be determined. While the Confederacy gained a tactical victory at Chancellorsville by virtue of General Lee's daring gamble, the battle was, in the end, a strategic defeat for the armies of the South. It was in this battle, after all, that Stonewall Jackson was accidentally killed by one of his own soldiers, who mistook him for Union cavalry. Applied to the *USS Chancellorsville*, that may mean what the digital theorist Chris Hables Gray has argued persuasively:

> Techs such as drones produce tactical successes, but strategic defeats in low intensity conflicts which aren't about kill boxes and rations but "winning hearts and minds." Something a drone cannot do.[5]

Or is it something else entirely? In the future, who can know with any certainty what will unfold with drones possessing subjectivity? Following science fiction, will drones rise up in merciless, mechanical revenge against their human creators? Or is something very different likely to happen? Will the story of affective drones repeat the lessons of human mythology because of which they were invented in the first instance and the memory of which will undoubtedly haunt them long after the disappearance of the human remainder? In other words, will the future epoch of drones, like the history of humanity before it, also be characterized by bouts of radiant, positivistic power mixed with accidents, futility, caprice, and the furies of enigmatic uncertainty?

Hydra Reawakened

Seemingly everywhere now, military visions are resurgent about weaponizing the oceans deep, about ensuring that advanced weapon systems are "proficient" in the blue waters of ocean depths, about gaining full military possession of the watery underworld.[6]

Is it possible that classical Greek mythology will finally find its practical realization in contemporary history by way of advanced military innovation? That, at least, is the hope of the US navy, as evidenced by the DARPA solicitation for innovative design proposals for a program named Hydra, which is aimed at creating permanent, unmanned underwater platforms in all the oceans of the world, populated by drones within drones. Media reports include enthusiastic accounts of "underwater drones deploying drones," "underwater drone carriers" for conquering the seven seas, and unmanned platforms for quick insertion of equally unmanned "air and underwater vehicles into operational environments." Here, the limits of nature are disappeared with rapturous speculation about US drone warfare moving from the sky to water, from surveillance from above to drone strikes from beneath deep water, shallow water, even from "river deltas or systems."[7]

Drones within drones, upward falling payloads, an unmanned undersea system: the future of drone warfare as envisioned by DARPA migrates the question of the unmanned from its previous station in targeted aerial surveillance to the depths of the seven seas. Here, it is no longer flocks of drones hovering in the sky, but something else – unmanned, underwater motherships equipped with drones within drones, some as troop transports, others as transport vehicles for armaments and supplies, all lying in wait, just offshore, just under the seas, waiting to instantly respond to insurrections, rebellions, disturbances.

While, from one viewpoint, this vision of repurposing the oceans for drone warfare provides another example of technological hubris combined with the US military's proclaimed ideological commitment to "global projection of power," from another perspective, it also contains a Heideggerian aporia. For Heidegger, the mobilization of the seven seas on behalf of a global system of command and control is part of the technical drive towards reducing nature and humanity to the status of the "standing reserve"[8] – the seven seas held in reserve, that is, for an innovative process of technological ordering with its "upwardly falling payloads," "drones within drones," and "underwater drone carriers."[9] However, there's one difference: almost as if perfectly symptomatic of profound, nagging anxiety about the eventual failure of the project in the face of a greater, as yet unknown, force, the very mythic name of the Hydra program announces in advance the most critical weakness of the initiative. After all, in classical Greek mythology, the figure of Hydra, this "serpent-like monster with two heads,"[10] evokes a larger mythological fable that is replete with moral complexity and martial ambiguity.

Mythically, the Hydra is always figured in relation to Heracles, the heroic representative of fallen divinity, who, in order to win back his immortality after killing his own wife and children, is forced to undertake twelve difficult labours, involving, among others, slaying the Nemean lion and capturing the Erymanthian boar, the Cretan bull, and Cerberus, with its three heads of a wild dog, tail of a dragon, and snakes emerging from its back. Yet, perhaps the most challenging of Heracles's tasks was overcoming the monstrous figure of the Hydra, who, with its ability to effortlessly regrow many new heads, guarded the swamps of Lerna, beneath which lay the entrance to the underworld.[11] The mythic force of the Hydra has to do with representing a fierce entanglement that only grows more difficult and complex with any and all attempts to overcome it. In the end, the Hydra was ultimately defeated by Heracles's brilliant tactic of cauterizing the severed heads one by one, thus eliminating the flow of blood and the generation of new heads. The lesson of the myth is that the Hydra, this watery defender of the underworld, is as weak as it is ferocious, an obstacle that can be overcome in practice by a skilled, creative, and courageous warrior such as Heracles.

Consequently, while the US military's Hydra program may well culminate in interesting designs for an underwater world populated by drones within drones, it offers no solution to the real problem, which is, in its essence, mythological – the always certain appearance of counter-power, of counter-resistance to sovereign claims of "ownership of the undersea domain" in the form of the new Heracles: a heroic figure – perhaps from the present, perhaps from the future – with a name of no importance and from a country of no significance, who can only win back political immortality by overcoming the new Hydra of the underwater drone.

Unfortunately, while the resolution of the problem of Heracles might have been left in suspension by an act of technological indifference, *naming* a project Hydra has about it all the signs of mythic necessity, posing a challenge to the sleeping powers of the long-neglected pagan gods of classical antiquity. Known now by the continuous appearance of the mythic signs of necessity, nemesis, hubris, and revenge, the spirits of those pagan gods have never really been at a distant remove from the technological scene and certainly have never been anything less than the essence, particularly if unrecognized, of political experience. While minds more attentive to the continuing mediation of the language of the pagan gods and the spectacular drives of technological hubris might enter a word of caution against carelessly conjuring up the forgotten spirit of the gods (particularly by formally inscribing their sacred names in the chronicles of contemporary history), it must be admitted that part of the unfolding truth of the most brash of the newest post-human gods – the language of technological mastery – is to issue a challenge to the death against the gods of classical antiquity.

Who knows really whether the power of Zeus, the jealous love of Hera, or the remorse of Heracles have heard the voice of this newest pretender to divinity? Like the original Hydra, this drone project sets out to guard the entrance to the underworld, no longer under the mythological swamps of Lerna, but within the watery labyrinth of the seven seas. Like the Hydra of classical mythology, this daring military innovation uses precisely the same tactic to propagate drones within drones – heads within heads – as a way of guarding itself against an enemy seeking to sever the only head of a multiplicity of heads that counts – the single, undetectable head of the Hydra that is immortal. Of course, in the transition from classical mythology to the new real of drone technology, the contemporary Hydra, lacking any plausible pretensions to immortality, begins the game of war with an immediate disadvantage. Now we know from the military's call for proposals that the underwater drone project is intended to operate in the real-time environment common to both the contemporary moment and, it should be noted, classical antiquity – insurrections, rebellions, civil strife, revenge-seeking suicide missions. Like the immortal Hydra, the underwater drone platform lies in wait, thus ceasing to be so much an instrument of spatial domination of the skies as a lethal weapon willing to engage *temporally* those who are courageous or perhaps foolish enough to gain entry to the underground. A time-biased technology operating in the liquid environment of the seven seas, this newest iteration of the myth of Hydra knows only that its weaponry of choice must, of necessity, be that of deception, subterfuge, and secrecy. Hiding in the depths of the oceans, revealing itself only when engaged in aggressive military strikes, the drones within drones that are the essence of the Hydra program adopt the language of temporality as their own: a waiting game of infinite patience with secret locations, illusions of identity, and hidden purposes. While drones hovering in the clear blue sky might communicate a message of terror by their very appearance, drones secreted within the seven seas communicate a different order of meaning altogether. Hydra is on the move again, with oceans bristling with hidden weaponry as the new deep blue sky of drone warfare.

The Drones of War

Perhaps nothing symbolizes so well the movement of power – from visibility to invisibility, from the imminent to the remote, from the language of discipline to that of a politics of control – than drone technology. Understood as a metaphor of power, drone technology represents the migration of power from something vested in the territorial claims of sovereign nations to the space-extending ambitions of trans-sovereign empires for which only the projection of power has political currency. Understood as a metonymy of power, drone technology is energized by the fact that, while it rides the imperial wave of the invisible, the remote, the monitor, its actual political effects are always highly visible, deadly intimate, and purely chaotic in terms of their impact upon targeted tribes, clans, families, communities, and individuals. Consequently, neither a pure metaphor nor an irredentist metonym, the power of drone technology rests structurally in its ultimately semiotic status as a violent, flickering signifier from the sky, an indeterminate point of mediation between invisible force and targeted visibility, remote commands and highly tactile results, unmanned control and social chaos.

Indeed, the fact that drone technology enters so easily and pervasively into contemporary public debate may be because there is something about the image of hovering drones – in all their invisibility, remoteness, and artificial control – that actually touches on, and is perhaps even emblematic of, an already widespread anxiety in the post-human condition. In this case, what we see in those images of Predator and Reaper drones in far-off lands, those almost post-apocalyptic scenes of violent power projected across the skin of the planet by way of electronic pulses sent from remote command and control locations, may actually bring to the surface of individual visibility what we already experience in our unconscious and more often than not unarticulated feelings as that which is most primal, and for that reason most uncertain, about the character of our necessarily shared political condition. Certainly, we can recognize that, as citizens of the privileged centres of neoliberalism, our political fate has for the most part already been structurally figured in advance. When imperial violence takes the form of unexpected and unpredictable blasts from the air, when imperial power depends for its very existence on subjecting dominated populations to a form of cynical power that operates like a murderous flickering signifier – invisible yet risible, remote yet intimate, controlling in its logic yet chaotic in its effects – then we too can recognize something not particularly alien to our experience yet deeply familiar. It is as if the massive deployment of drone technology by the permanent war machine represents an accelerated test bed for a new form of political ethics yet to emerge, one deeply attuned to the language of weapons of invisibility, death matrices by remote command, and power at that point where it becomes something less terrestrial than purely atmospheric, something as intimately present as it is technologically suffocating.

When the drones of war are tested in foreign lands, we can perhaps comfort ourselves with the moral illusion that politics today neatly divides into a more primary distinction between friends and enemies, and that the boundary points for such a division can be identified by the signs of citizenship, religion, ethnicity, and race. While such ready-to-hand distinctions have the grisly political benefit of ethically dividing the world into a sacrificial table of values upon which will soon be arranged those to be violated as the unlawful alien, the scapegoat, the terrorist, the enemy non-combatant, or the stranger, they also have the strong moral appeal of rendering any and all violations of the norms of social justice ethically justifiable. Those not structurally determined in advance as the outsider – from the alien to the stranger – will probably never know what it means to inhabit a body, a race, a family, a clan, a tribe, or a society that will never be honoured with the most elementary rights of human recognition and reciprocity – the right to be mourned, the right to grieve. While media communiqués about new drone strikes in Asia and Africa are usually figured in the deliberately sanitized and entirely nebulous language of the War on Terror, these reports on the conditions of the new security state often do not provoke even a ripple of discontent precisely because they work to confirm an ethically striated vision of the contemporary political condition that we long ago interiorized as our own. Of course, having acquiesced either consciously or by a silent proxy in the privileges to be gained by linking our fate to the ethical exclusions necessary to the self-preservation of power, we find there remains just the tangible hint of a doubt that someday the moral cycle of accidental divisions and ethical cynicism represented in all its ferocity by the drones of war will run its full course.

Perhaps they already have. Perhaps the political use of drone technology to terrorize often defenceless populations rests on a prior moral blast that has already obliterated much of the traditional language of human reciprocity and recognition. More than we may suspect, we are already dreaming with drones. Dazzled by this spectacular projection of technology that advances the space control of empire against time-bound forms of terrestrial resistance and perhaps, in consequence, numbed by the silent ethical compact that authorizes lethal violence from the air, we may have already naturalized something resembling drone subjectivity as our very own interior habitus. But if this were the case, then it would only be churlish to later claim that we did not have at least a premonition of our own approaching extinction event when the drones come home. In this sense, what is actually being tested in far-off foreign territories may not be, in the end, the purely technical abilities of drone technology as instruments of war but pilot projects for the use of drones at home. When power turns inward, as it always does, when that which has been done by power to those determined to be beyond the rites of grieving and mourning finally turns on us as the new ungrievable, the definitely unmournable, we should not be surprised. As the last and best of all the cynical signs, drone subjects have long been nurtured in the language of moral equivocation: subjects of use and abuse, subjects of control and chaos, subjects remote even to themselves in their most intimate moments. With this perfectly equivocal result, when drone technology tracks back to its country of (technological) origin, when drones become an important dimension of the language of the new real, what the consequences will be is still unknown, still emerging. Yet we do know this: the contemporary situation oscillates today between scenes of "Bounty on Drones" and the present and future spectre of "Drones Hunting Humans."

Bounty on Drones

We are now all too familiar with drones hunting humans – that's the essence of drone warfare – but what about the opposite, humans hunting drones? There was a media report about an ordinance proposed by the rural township of Deer Field, Colorado, not only licensing hunting season on passing drones but also offering a bounty for confirmed kills. This report was quickly followed by full government panic, with the Federal Aviation Administration (FAA) issuing a formal cease and desist warning.[12]

So then, we have a perfect reincarnation of the spirit of the Wild West in the early years of the post-human condition. Not settling for legal niceties and certainly not yielding quietly to official power, some citizens of Deer Field want to do what gun-toting trailblazers of the old West have always done before them: take to the new surveillance trails of the sky in order to bag a hovering drone. This movement leads to the question, What happens when the drones finally do come home? What happens if they come home, not as super-tech augmented, sky-bound survivors of hard-fought battles in Afghanistan, Somalia, and Yemen but as drones making their first appearance in their homeland as a new form of heightened state security – only this time, a state security framed less as a scattering of insurgent rebellions in the far-off global reaches of the empire of neoliberalism than as returning drones pressed into service again as the front (aerial) line in the surveillance of domestic populations? Will this be the first symptomatic sign that the power of the new security state, having fine-tuned its apparatus of control in the War on Terror, is finally prepared to colonize its own population?

Or perhaps the drones come as something else: not the new security state filled with drones solely in passive submission to its strategic aims, but a present and future time populated by drones replete with a multiplicity of commercial purposes – unmanned media photography; drones for the surveillance of wildlife, crops, and storm damage; long-distance rescue drones in otherwise inaccessible harsh environments; knowledge-based drones for long-distance, real-time education. While the commercial repurposing of drones is indeterminate – limited at first really only by the human imagination – one thing is certain: when the drones come home, the future is likely to be stellar gridlock, with the sky the limit for the sudden extension of human commerce. It is likely, then, to be a future filled with many accidents in the air, many near collisions, as drones speeding along on their different missions forget to keep watch on their unmanned neighbours, and of course filled with the inevitable – long lines of stalled, sky-bound traffic, impatiently spinning their algorithmic wheels, probably getting into hot coded disputes with quick terminations of flight high on the probability charts. In this scenario, the skies of tomorrow are the expressways of today: studies in immobility with patience at the limits of its endurance.

Yet, there may be this difference: taking advances in ubiquitous computing and relational processing to its extremes, there are undoubtedly technological plans now afoot to design drones of the future – or, as the drone industry likes to call them, "unmanned aerial vehicles" – with the capability of communicating with one another, sharing pertinent information, and what's even better, with advanced capabilities to mimic bees and birds by instantly gathering into fast-moving flocks of flying drones. No longer, then, would we have drones as long-distance, lonely fliers, but suddenly there are drones armed with hive-minds for better swarming, flocking together in the air, swooping down on unsuspecting drone singletons, and just as suddenly turning every which way probably for the pure joy of being swarm. For most of us civilians who are aware of brilliantly creative technological advances set in motion without much thought given to unexpected consequences, it is unlikely that computer console designers of swarming drones have read up much on insect lore, the fact that swarms in nature appear in many different shapes, all of which have very real-world consequences – from the hard-working cooperatives of bee hives with their built-in aristocratic class structure of drones and queens to the hornets who form angry swarms when annoyed or angered by the human presence. So, while the utopian dreams of all the drone designers probably shade away into the comforting flight path of future unmanned aerial vehicles as busy bee hives in the sky, the hard reality will probably be something very different. Since drones first came into existence at the behest of military violence, with its calculated bursts of murderous rage, there's no reason to think that future generations of commercial drones will not, at some undefined point, rekindle memories of the targeting imagination of the Predator and the killer instinct of the Reaper as the most active, and fondly thought of, long-term memories.

When drones themselves begin to dream, their psychoanalytical drivers will probably move unerringly to that moment when drones as purely technological devices first merged with human psychosis on many battlefields of the past. It won't move there accidentally, but with a deliberate and almost inevitable evolutionary logic, since the killer instinct, with all its preparatory conditions – surveillance of targeted populations, data acquisition, arming of weaponry, and bursts of destructive violence – is not at second-hand remove from the logic of drones, but actually designed into their unmanned (but not unarmed) intelligence. The curious mixture of cybernetic rationality and spasms of irrational violence has always been the emblematic sign of drone wars. Like the human species before it, drones also have memories. Sometimes these memories are short term, like agricultural fields to be surveyed, packages to be delivered, isolated survivors to be rescued, but they can also be long term. It is those deeply embedded, long-term memories of their all-too-human origins in a mythic mix of antiseptic designer rationality and murder from the air that will most likely be activated by swarms of drones. Liberated intentionally from human control with sensors fast processing the territory below and other drones alongside, the drone swarm, like those angry hornets before them, is a likely candidate to go on instinctual killer mode, to become, in practice, what their drone ancestors had long ago initiated in the skies of foreign lands. In this case, when the will to technology is finally realized in the form of swarms of angry drones, when cybernetic reason merges with unmanned violence, there will probably be a big rush for those hunting permits.

When drones become an unmanned aerial species, equipped with autonomous intelligence, weapons of choice, surveillance capabilities, and laser-like targeting abilities, we will probably be able to discern that their primal psychology will not run to the hectoring superego or the reasonable ego but to the instinctive-like drives of the howling id. Without the disciplinary cage of the social to tame it, without the fear of God to inhibit it, drones of the future will make their first swarming appearance as the id unbound: psychically self-possessed, humourlessly destructive, seemingly irrational but, for all that, cunning, creative, and probably (cybernetically) ruthless in the games they play.

With this in mind, the unsuspecting residents of Deer Field might be wise to start running or, at least, to instruct their children in the ways of mythic nemesis likely to be expressed in their streets when drones appear in the domestic sky.

Drones Hunting Humans

The first of all the violent invasions of drones hunting humans has already taken place. That's the so-called War on Terror, with its carefully orchestrated publicity campaigns in support of ever-increasing popular fear and, in a perfect feat of logical symmetry, its identification by means of the US administration's "disposition matrix" of a changing list of "terrorists," some perhaps even dangerous opponents for targeting by fleets of drones stationed in the skies of designated kill zones. For example, according to media reports from the tribal areas of Pakistan and adjoining regions of Afghanistan, we can gather some preliminary results of this lengthy experiment in test-driving drones hunting humans.[13] Indeed, similar to large-scale, innovative scientific projects that can only seek major funding on the basis of "proof of concept" projects, the War on Terror might be viewed retrospectively in the same terms. Here, all the design ingredients were mobilized for a potentially successful "proof of concept" experiment in drones hunting humans: a captive population that can be targeted at will; the necessary long-range territorial distance (from Las Vegas to Afghanistan) needed to field test lag time for the remote control of unmanned weapons systems; and media mobilization of the public opinion of domestic populations, which generates active support for the frequent use of unmanned aerial vehicles in warfare but, more importantly, generalized ethical tolerance for excluding targeted populations, whether targeted "terrorists" or civilian bystanders – families and friends at funeral gatherings, children sleeping at home – from basic recognition of the rights of reciprocity as human beings. From this purely strategic point of view, the experiment in drones hunting humans, which was the essence of the War on Terror, was demonstrably successful. Not particularly, of course, was the deemed success based in the numbers of known "enemy combatants" killed – it was always the usual folly of war to expect that seasoned warriors adept in the ways of camouflage and surreptitious movements could be tracked, let alone eliminated, by flying robots in the sky. But in the usual way of things, even major failures like the rash and ill-conceived military adventure in Afghanistan have their purposes. Like any cold-eyed examination of the outcome of this proof of concept experiment, the results were strikingly successful in inverse proportion to the harsh reality of the overall military failure itself: fleets of Predator and Reaper drones could be controlled remotely; brilliantly displayed, real-time videos of actual combat situations could be provided to elite commanders bunkered down in the command and control centres of the Pentagon, intelligence services, and the White House itself; captive populations could not only be targeted as required but, as an added benefit, future psych-ops would be guided by the medical finding that the humming presence of drones hunting humans in the sky would accelerate mass psychological depression, and thus political paralysis, in the targeted population; and finally, domestic populations have quickly and decisively proven themselves receptive to, if not eager participants in, ethical indifference to those identified by the state as fit objects of sacrificial violence. Consequently, when drones first began to hunt humans in the War on Terror, a complicated calculus of proof of concept was affirmed, one that was at once strictly *technological* (remote control of unmanned aerial vehicles), *specular* (those live video feeds to the masters of the war machine providing, at the minimum, the illusion of being warriors, if only ersatz warriors, in games of life and death);[14] *psychological* (creating and maintaining a generalized condition of cultural acedia in targeted populations); and *ethical* (preserving political support for drones hunting humans by intensifying that sweet spot of all carefully orchestrated military media campaigns – a perfect blending of moral indifference mixed with feelings of righteous anger as the emotional fuel supporting war drones operating under the sign of abuse value). This, in effect, constitutes the technological ontology of surveillance practices that function as the operating system of the new security state.

Now that the "proof of concept" stage for drones hunting humans has been completed, it will only take a slight redesign of contemporary models of war to successfully re-enact this very same mix of tactics, logistics, ethics, and psychological animus in domestic space. Following the doubled ideological logic of facilitation and control by which new technologies are usually introduced, we can already identify the key political markers facilitating drones hunting humans at home. Not surprisingly, everything will have to do with "securitizing the homeland." It will not be just securitizing the always porous borders in the face of increasingly phantasmagorical anxieties about "illegal aliens" and sometimes even legitimated suspicions about potential terrorist attacks, but also the much-publicized need to securitize dense networks of oil and gas pipelines, isolated power stations, nuclear facilities, and transportation corridors. In this case, when the drones come home, it is likely that the form of invisible surveillance will take over the open skies of homeland security, with "upmoded" warlike drones securitizing borders, patrolling far-flung networks of pipelines, and surveilling over targeted cities, neighbourhoods, homes, vehicles, and individuals. While economic insecurity and political anxiety provide, in the first instance, the necessary conditions for authorizing the apparatus of drones hunting humans into the domestic scene, the future will be different.

Art as a Counter-Gradient to Drone Warfare

When machines break the skin's surface, becoming deeply entangled with desires, imagination, and dreams, do we really think that we will be left untouched, that easily discernible divisions will remain among the machinic, the natural, and the human? Without conscious decision or public debate, we may have already passed into the deeply enigmatic territory of the new real: that space where the price to be paid for the sudden technological extensions of the human sensorium may be an abrupt eclipse of traditional expressions of consciousness and ethics; that time in which the uniform real time of Big Data effortlessly substitutes itself for the always complex, necessarily enigmatic, and lived time of human duration. When the human life cycle increasingly depends for its very existence on technological resuscitation, how much longer will the meaning of the human not yield to the greater power of the technological? That's the new real: the future world that is now, where individuality singularity has been replaced by network connectivity; where bodies of flesh, blood, and bone have already been surpassed by a proliferation of electronic bodies in the clouds; where every step, every breath, every glance, every communication gives off dense clouds of information that are, at once, our permanently monitored past and our trackable future. For some, it is definitely suffocating. For others, it is a fully liberated future of the transhuman, where the handshake made between the codes of technology and the missteps of humanity indicates that we have already migrated into another country, another time, with sublime possibilities for technologically augmented bodies, digitally enhanced vision, and quickly evolving light-wave brains.

We have always been an adventurous species, living at the edge of dangerous risks and practical wisdom, a species (technologically) willing to will its own extinction while, at the same time, artistically probing the future for its terminal abysses and points of creative transformation. The unfolding story of drones is the very same. It is the *artistic imagination* of drones that displays heightened sensitivity to what Heidegger might have described as the new dwelling place of drones at home and drones at war. Refusing to think outside the imaginary landscape of drone technology, the artistic imaginations can be so replete with important insight because they actually engage in the material reality of drone technology. Their engagement is not through active imitation or complacent praise; rather, they are artistic imaginations that think right through all the symptomatic signs of drone technology to discover its essence – not only that which is made visible by drones but how its very invisibility and remoteness burrows inside human anxieties.

Today, a number of contemporary artists act as leading political theorists of drone technology, exploring in the language of aesthetics the remote violence and the equally remote ethical distancing that occurs when unmanned aerial vehicles are purposed by larger military missions. In the contemporary artistic imaginations are to be discovered the full dimensions of drone technology as the truly ominous symbol of the times in which we live: *a symbol of power* that is remote, invisible, and weaponized. Representing, in effect, heightened cultural consciousness concerning the full implications of drones, artists often function today as the kind of philosophical explorers that Hannah Arendt once described as the "negative will"[15] at the heart of technology: a pornography of power that seeks to draw everything into obscene visibility – desensitized, dehumanized, sadistic in its pleasures, cynical in its purposes. Opposing the secrecy that surrounds the development and application of militarily purposed drone technology, contemporary drone art – online and real time – breaches boundaries of secrecy by making its aesthetic explorations fully open to the electronic public, linking together in common ethical purposes drone artists from different countries and, perhaps of greater significance, creating active collaborations between critical drone art and the actual and potential victims of the cold violence of those unmanned aerial vehicles, hovering in the skies of foreign lands for the moment and soon in the twilight sky of the imperial homeland.

#NotABugSplat

In military slang, Predator drone operators often refer to kills as "bug splats," since viewing the body through a grainy video image gives the sense of an insect being crushed.

– #NotABugSplat[16]

#NotABugSplat, an emotionally evocative and deeply ethical project by a Pakistani artist collective, is what happens when those held under the sign of erasure by warlike drones finally have the opportunity to speak publicly and in doing so begin to imagine another language, ethics, and memory for making the invisible visible, the prohibited image the necessary subject of moral inclusion, and the (technically) silenced a suddenly noticeable, deeply insistent subject struggling to be recognized. When the governing ethics of power privileges a form of long-distance ethics essentially constituted by a strict separation between decision and consequences, between remote drone operators and slaughtered people in fields, then we can most definitely know that ours is a culture which moves at the ethical speed of a bug splat with all that that entails in terms of extremes of dehumanization, desensitization, and pure objectification.

Understanding that the only effective ethical response to power under the sign of a bug splat is one that suddenly humanizes the field of remote vision and thereby activates an insistent demand for recognition as human beings, #NotABugSplat works to facialize Pakistani victims, actual and intended, of US drone strikes in order to make legible the human dimensions of those condemned to abuse value status in the age of drone technology. The artistic strategy is as straightforward as it is ethically profound.

The image released as part of this project was taken by a mini-helicopter drone and depicts a young girl who lost both her parents in a drone strike in Pakistan's Khyber Pakhtunkwala province. Hoping to instil "empathy and introspection," one of the artists of the organizing collective said: "We tried to replicate as much as we could what a camera from above will see looking down ... [W]e wanted to highlight the distance between what a human being looks like when they are just a little dot versus a big face."[17]

While the artistic project involves, in the first instance, remaking a farmer's field in rural Pakistan into a large art installation featuring a massive image of a young girl's face – an image aimed at activating the ethics of remote predator drone operators – the political implications of #NotABugSplat are universal. Here, in a unique case of art acting as a counter-gradient to power, that haunting image of a young Pakistani girl "who lost both her parents and two young siblings in a drone attack" reverses the language of power by critically and decisively re-ordering the logic of targeting. Until this point, the specific targeting of drone attacks was solely a matter of cold military logic with, for example, all young males in strike zones considered "militants, unless there is clear evidence to the contrary," and the local population deemed "guilty by association" and "a militant if they are seen in the company or in the association of a terrorist operative."[18] Working to undermine the antiseptic, radically indiscriminate logic of "signature strikes" with their unreported but widely documented massive civilian casualties, #NotABugSplat subverts such a logic of targeting. While it might be naive to suppose that an image, even a large haunting image, visible to predator drones would have any real effect on the ethics of their remote operators, this attempt to make suffering visible, to actually facialize those literally objectified by technologies of violent disappearance, has an unpredictable advantage. For the very first time, the ethical worm turns by a radical reversal in the order of targeting. Suddenly, an art installation in a rural, Pakistani field begins to speak to drone operators housed in the remote reaches of an imperial homeland, targeting their ethics, their memories, their most fundamental understanding of the necessary demands implied by human recognition and reciprocity. While the nihilism evinced by drone technology may already be so advanced as to immediately nullify the ethical purposes of the artistic project, there always exists the fragile, nebulous possibility that the face of existential suffering can give pause to the most arid, most unmanned of technologies of contemporary war. In this case, #NotABugSplat might best be viewed as the first of all the future artistic experiments in breaking not the sound barrier of earlier times but the ethics barrier of remote technology. Consequently, it is in this emotionally compelling project – a project that puts the question directly concerning whether or not shared ethical responsibility can triumph over the singular purposes of drone warfare – where both the last and best hopes of suffering humanity surely rest.

Terror from Above

Let me tell you a story
   a bedtime story
Let me tell you a story
   of Predator drones with giant wings
   equipped with hellfire missiles
   and "light of God" lasers
   choking the skies over northwest Pakistan
Let me tell you a story
   a daytime/nightmare story
   of grandmothers as "bug splats"
   and children as "double taps"
Let me tell you a story
   an everyday story
   of terror from above
   villagers burned, body parts strewn
   over cultivated fields
Let me tell you another story
   the official story
   a drone warfare story
Let me tell you a story
   of precision strikes
   where no innocent is mutilated, incinerated
   or murdered
Let me tell you a story
   But we know *this* story is a lie

Surveillance power increasingly functions by moving from the centre of human attention to its peripheries – invisible, ubiquitous, waiting. Now it is no longer a matter of people having to walk into the field of machinic vision – as it was in the age of street-level video cameras – but of a machinery of surveillance that electronically scans entire landscapes, carefully monitoring the daily habits of their inhabitants, watching for selected disturbances of the field of vision that may potentially trigger a violent technological reaction – a drone strike. In this case, the surveillance power of drone technology is no longer limited to a list of potential targets listed on what the National Security Council describes as the "disposition matrix" but something more menacing, namely the harvesting of entire populations under the sign of a *generalized disposition matrix* – people who are deemed to be in a permanent state of suspicion by associations no matter how accidental, by physical proximity through a wedding, a funeral, a community gathering, by the simple geospatial fact of where they happen to live. When surveillance migrates from visible technologies to invisibility, from reliance on human disturbances of machinic vision to machinic disturbances of individual experience, it means that we are living in the era of space-binding power – always hovering on the peripheries of life, bracketing the lived time of those inhabitants held under suspicion by the prospect of an immediate sentence of death from the air. What does it mean, then, when the power of surveillance is no longer limited to visual scans of always threatening populations but itself incorporates a politics of life and death? Equally, what is meant when entire theatres of war in the contemporary era themselves retreat behind a shield of invisibility: unreported, unexamined, undisturbed? What, in effect, is implied by the present state of affairs, when the concept of invisibility itself has been weaponized? While technologically augmented society likes to pride itself on the culture of connectivity, with bodies everywhere seemingly globally mobilized by social media into always open data points, the reality of the new invisibility associated with technologies of surveillance would intimate that, in some fundamental sense, we are actually radically disconnected from some very essential knowledge. Perhaps what we are most disconnected from is the sudden transformation of weaponized invisibility – surveillance technology in the form of drone strikes – into a key expression of the ontology of the times in which we live: drone strikes as being towards death.

The political implications of drone strikes as weaponized invisibility have been brilliantly explored in the aesthetic work of the British artist James Bridle. In an interview with the BBC, Bridle noted that his art is interested "in exposing the connection between secret surveillance, power projection, and new technology through installations."[19] With an artistic imagination that explores the contradictions and strangeness of contemporary society, Bridle argues that, while our phones allow us to see the world at a click of a mobile button, at the same time we generally have little knowledge about the actual battlefields on which wars are being waged. His overriding aesthetic purpose is "to use these technologies to make visible the contemporary battlefields, these drone strikes."[20]

Working in the language of social media, one of Bridle's aesthetic projects – *Dronestagram* – repurposes Google Earth into a visual cartography of actual drone strikes, including location, frequency, and timing, that is then circulated through the electronic capillaries of social media, from Instagram to Twitter. Here, one medium of (social) communica-tion is creatively redeployed as a way of drawing into visibility another medium of social destruction. Beyond *Dronestagram*, Bridle has initiated another interesting project, one that has a larger collective purpose: to create public awareness of the material reality of drone strikes. Titled *Drone Shadows*, this project, based on the active collaboration between Bridle and Norwegian visual artist Einar Sneve Martinussen, produces perfectly scaled chalk drawings of drone shadows in the streets of many cities of the world. Bridle states: "One way of looking at drones is as a natural extension of the internet ... in terms of allowing sight and vision at a distance. They're avatars of the net for me."[21] Or, as one insightful commentator has noted: "In Drone Shadows, he draws a chalk outline to scale of a different drone each time, highlighting that not only do they not cast shadows from the vast height they operate at but that they are here among us, very literally, and unseen."[22] In a larger sense, Bri-dle's overall project, what he describes as the "New Aesthetic" – whether *Drone Shadows* or *Dronestagram* – focuses on the complex entanglement of technology and warfare as the essence of invisibility itself. By creat-ing shadows for that which is without shadows, by visually mapping that which wishes to remain unmapped, his artistic imagination probes the full consequence of invisibility itself. In so doing, the project renders the question of invisibility even more complex in another way. While drone strikes can be mapped and drones themselves made to cast chalk-like shadows on city streets, what about those other invisibilities, those growing invisibilities of language, culture, ethnicity, geographical loca-tion – of life itself? Why is it that so much of what is visible today is, in fact, invisible? Why is it, in the end, that only certain expressions of human visibility – targeted bodies in the tribal lands of Pakistan, Yemen, Somalia – are dragged out into the violent visibility of otherwise invisible technologies of surveillance? Have we reached a first cultural, and then political, breaking point in which the meanings of visibility and invisibil-ity have entered into a more complicated mediation, one in which the question of visibility will increasingly rely on a greater political ordina-tion, while those other very human invisibilities – differences of class, race, ethnicity, life itself – are allowed to disappear into the category of human remainder? Of course, there is also this curious, purely aesthetic paradox, namely, that the act of making visible those hidden warfare in-visibilities of Predators, Reapers, and Global Hawks does not rely on any-thing particularly high tech but on two other expressions of more urgent technologies – the simple act of drawing chalk outlines of drones on city streets and the very public act of mobilizing global public participation in the art of making drones visible.[23]

Night Sky Epilogue

The night sky drone
   is a bullet, an eye, a gut
   spilling blood
Venus transits and the sun
   is a distant memory
2 tons of fuel and a ton of
   munitions. 18" and 7,000 miles
Palm trees
   the smell of BBQ
   surfers, scubas
   walkers and runners
A biplane overhead laconically
   pulls a sign that reads
    "There's no place like home
    especially when it is clean and green."

4 Robots Trekking across the Uncanny Valley

Blended Reality

In the new real, we are running with the robots: industrial robots for seamlessly automated car manufacturing; medical robots for facilitating patient care in assisted living retirement communities; warrior robots engaged in materializing the imaginative game scenarios of cyber warfare; toy robots that promise a happy first encounter between machines and the newest generation of humans; and, most of all, invisible robots circulating in the data clouds of social media as sociobots. Perhaps more than we may suspect, ours is already a blended reality in which robots not only live among us as artificially programmed prosthetics equipped with articulated limbs and complex sensory arrays but have also begun to live *within* us, quietly but insistently bending the trajectory of human perception, imagination, and desire in the direction of a future life of the mind that bears unmistakable signs of a robotic imaginary. Consider, for example, the following stories focusing on the complex intersection between human intelligibility and robots, both invisible and visible.

Neurobots

While the future of human encounters with robots has often been envisioned as an ominous struggle between fragile, but immensely adaptive, humans and powerful, although less creative, mega-robots, the real-world encounter has proven to be decidedly low key, ubiquitous, and technologically subtle. Seemingly everywhere, the digital body has been swiftly delivered to its robotic future in the form of a pervasive network of invisible bots: *sociobots* swarming social media sites creating contagious flows of viral information, influencing individual perception, imitating human behaviour; *capitalist superbots* in the form of high-frequency trading algorithms that powerfully shape the ebbs and flows of stock transactions; *psy-op bots* in the service of military intelligence that function to effectively influence political perception; and, of course, those other multiplicities of net bots – *spiders, crawlers, and malware* – that trawl the internet, sometimes like proletarian worker robots performing routine web indexing functions but at other times like futurist versions of the Cylons in *Battlestar Galactica,* quietly searching for critical weaknesses in websites, software programs, and internet infrastructure itself. Consequently, to the question concerning future encounters between humans and robots, the answer is not only already well known but also pervasively experienced as the contemporary real-time environment of digital life. No longer content to remain at a safe, mechanical distance from their human creators, robots in the form of those lines of code that we call "bots" have already broken down the walls of human perception, inhabiting the world of social media as their cybernetic hive, attaching themselves to the human imagination in the seductive form of hashtags and tweets, and, all the while, migrating the spearhead of robotic evolution itself from the mechanical to the neurological.

In the usual way of things, no one really anticipated that robots would faithfully follow the trajectory of technology itself, from high visibility to pervasive invisibility, travelling from the outside of the human body to the deepest interior of human subjectivity, quickly evolving from the mechanical to bots with very active cognition. When bots proliferate in the digital clouds that surround us, when they actually take up neurological residence in human perception, desire, and imagination, we can acknowledge with some confidence not only that we are already running with the robots but that something more uncanny is taking place: namely, that robots are already living among us and, most decidedly, living *within* us. The meaning of this is fully enigmatic. When robots were something that we could see – for example, the cute Japanese robot that played soccer with an American president and concluded with a victory dance and cheer[1] – we could take the measure of the event in traditional humanist terms. But what happens when robots actually trek across the uncanny valley? We don't mean "uncanny" in the usual sense of the term, because robots physically start to become indistinguishable from humans, but in the deeper sense that bots are perhaps already an indispensable dimension of post-human subjectivity. We mean it literally. For example, it is reported that 30 percent of all Twitter content is non-human (bots and algorithms): bots replying to articles, bots masquerading as friends in order to redirect traffic, bots for spying, for trading, for porn: spambots, bimbots, sociobots. As one media report[2] notes, sociobots are amazingly adaptive "automated charlatans" programmed to tweet and retweet, supported by vast databases all the better to fool their human interlocutors, "fake sleep cycles" to better imitate the slow rhythms of natural life, and replete with "quirks, life histories and the gift of gab." Their presence is felt everywhere: advertising on social media, swaying elections, trading stocks, phishing, sparking scandals, rumours, suspicions, and anxiety.[3] The report goes on to note that half of net traffic is non-human and that, in "two years, about 10 percent of the activity occurring on social online networks will be masquerading bots."[4] In that case, we are already deep into the strange times of a fully blended reality of humans and non-humans with the urgent question remaining, Have we become our own uncanny valley?

More than the sheer quantity of sociobots invading every dimension of digital life, what is significant about this report is something left undisclosed: that bots are integral to the question of social identity. We don't simply mean "integral" in the sense of leveraging perceptions, desires, and imagination to move in certain directions but "integral" in the fuller sense of the term – that, perhaps, we have already succeeded in moving beyond the point of real-time familiarity with the presence of bots to actually being part human/part bot. In this case, what may be truly uncanny is our own online subjectivity, occupying as it does an entirely unstable boundary between lines of code and lines of skin. When bots come inside us, pacing our existence with their artificial "sleep-wake cycles," mirroring our moods with "persona management software," and creating networks of their own consisting of "friends and like-minded followers," we can recognize that we have become the first and best of all the post-human subjects, breathing in lines of code as the real source of digital energy that allows us finally to come alive as the flesh and blood of sociobots.

More than half a century ago, the American psychologist B.F. Skinner correctly (and in fact enthusiastically) endorsed a future society based on a relatively primitive theory of "radical behaviourism."[5] Setting aside enduring questions concerning the origin and meaning of introspection and unconscious desires, Skinner suggested an alternative form of human subjectivity constructed on the strictly behavioural foundations of "operant conditioning."[6] For Skinner, what matters is the *quantified self*: the observable self that acts in and upon the world on the entirely predictable basis of social reinforcements – some negative (punishment), others positive (rewards), with yet still others more neutral in their role as reinforcements. Reducing the diverse spectrum of individual human experience – lingering desires, upstart passions of the heart, long-buried psychological repressions, mixed motives – to the observable behaviour of a subject that is postulated as acting on the basis of a social protocol of rewards and punishments (avoiding that which hurts, privileging that which rewards), Skinner's vision held that what was true in the laboratory with respect to the behaviour of rats and pigeons was equally true of social behaviour in general. In other words, human behaviour could actually be modified by the application of the soft power of a token economy, providing actual, and sometimes symbolic, rewards as an inducement for certain privileged forms of social behaviour, while gradually extinguishing undesirable behaviour by the hard power of pain and punishment. Stated in its essential elements, Skinner's vision of social behaviour – "operant conditioning" – provided a way of transcending millennia of concern with that strange and definitely precarious mixture of animality, intellectuality, and emotion that is the nature of being human in favour of an ecstatic theory of remaking humans by the organized application of a radically new technology of human subjectivity – *radical behaviourism*. In this perhaps pragmatic and certainly deeply visionary theory of the human condition, there was always a twofold ontological assumption: first, that persistent concerns with supposed epiphenomena such as psychic blockages, unknown motives, and interior sensibility could, and should be, dismissed in favour of a technological vision of subjectivity open to its surrounding environment, deeply influenced by its actions, and responding accordingly; and second, that the "self" of radical behaviourism could be socially modified, indeed socially engineered, by the methodical application of the principles of operant conditioning. Curiously, while at the intellectual level, the technological utopia that Skinner envisioned in his books *Walden Two*[7] and *Beyond Freedom and Dignity*[8] was surpassed by theoretical debate about the rise and decline of all the referentials of truth, power, and sexuality, Skinner's prophetic vision of a social self capable of being modified by the soft power of social reinforcements – particularly the "token economy" of radical behaviourism – has finally found its key public expression in the once and future society of sociobots. Not simply a new technology of communication perfectly fit for the age of social media, sociobots are, in their essence, something very different, namely a technology for modifying human subjectivity that is simultaneously political and neurological: *political* because sociobots embody how the ideology of operant conditioning is inserted into the deepest recesses of the data mind – the externalized, circulating consciousness characteristic of the quantified self of social media; *neurological* because sociobots are the primary cybernetic agents of "cognitive hacking," that complex process whereby the key driver of the newly emergent attention economy – perceptual attention – is encouraged to turn in certain directions, sometimes by positive reinforcers operating in the language of seduction and at other times by negative reinforcers functioning in terms of fear and anxiety. When swarms of sociobots attach themselves to the data mind – flirting, chatting, spying, tracking – we can clearly recognize that we are already living in a society of soft power and modulated violence.

Indeed, one of B.F. Skinner's most celebrated instruments for test-driving the theory of operant conditioning was the "Skinner Box,"[9] a closed, programmable environment whereby test subjects – including laboratory rats and pigeons – could be probed, reinforced, and, if necessary, punished as a way of calibrating, and thus engineering, the protocols of effective social modification. Now, just because Skinner's theory of operant conditioning – with its stripped-down assessment of human behaviour, its studious attention to the best practices of a token economy, and its transcendent vision of behavioural modification guided by experts – was seemingly displaced by theoretical attention to the death of the subject, from poststructuralism and postmodernism to post-humanism, and most recently by new materialist theories focused on the complexity of objects as life forms, does not necessarily mean that operant conditioning, with its profoundly eschatological vision of behavioural modification, was lost to the world of emergent technologies. In one of those superb ironies of cultural reflection, the Skinner Box could be quickly left behind as so much detritus on the way to post-human culture precisely because the theory of operant conditioning was always waiting patiently and persistently for its technological realization by a creative form of new media – in fact, social media – that could instantly and decisively translate the anticipatory vision of soft power, token economy, and reinforcement theory that was the Skinner Box into the generalized network of sociobots within which we find ourselves enmeshed today. In this case, when sociobots take active possession of social media, when complex patterns of human neurology expressed by the ablated consciousness of the data mind are gradually shaped, indeed modified, in their observable outcomes by bots that chat, make suggestions, anticipate connections, manifest seemingly total recall, and facilitate the attainment of desirable goals (better health, greater intelligence, early warnings), then, at that point, the Skinner Box is no longer an object outside ourselves but something else entirely – a technology of programmable subjectivity rendered part flesh/part data. Today, it is not so much that we are mingling with physical robots in ways anticipated by cinematic and science fiction visions of the technological future, but that clear, discernible borders have been eliminated between immaterial (social) robots and ourselves, so that it is difficult to know with any certainty whether a friend or a commentator on social media is human or the sensitively attuned response of an artificial life form – a sociobot – who can know us so intimately because, in daring to become fully digital – *being* social media – we may have inadvertently entered in the long-anticipated world of B.F. Skinner redux. Replete with swarms of bots – sociobots, neurobots, spybots, junkbots, hackerbots – the ablated Skinner Box that is the universe of contemporary social media has this common feature: expert systems in the form of artificial life forms function ceaselessly to modify, cajole, influence, and channel the privileged psychic targets of human perception and social attention in the token economy of network culture, with its powerful technologies of soft facilitation and its equally harsh technologies of command, including surveillance and tracking. Happily taking up neurological residence in the data mind, armies of neurobots, sometimes acting at the behest of corporate capitalism or perhaps under governmental supervision, are, in effect, the way in which power speaks today – otherwise invisible databases that seduce, inform, link, and recall as leading spearheads of evocative communication between robots and humans.

With the sheer invisibility of sociobots, the fact that the first, fateful encounter between robots and ourselves occurs in the innocuous, immaterial form of lines of code may intimate the elimination of the pervasive anxiety surrounding the "uncanny valley" – that psychic moment identified by robotics engineers when robots are effectively indistinguishable from human presences. In this case, the uncanny valley of robotics engineering lore may well constitute an ancient, psychological reinforcer supporting the pattern maintenance of established boundary lines long viewed as necessary to the self-preservation of the human species. While lines of code never rise to the psychological prominence of increasingly human-like mechanical robots, they do enjoy an important technological attribute, namely encouraging the human species, individually and collectively, to drop its traditional psychological aversion to mixing robotic and human species identity, which thus increases the vulnerability of the human species to quick insertions of the most fundamental elements of robotic consciousness such as ambient awareness, distributive consciousness, circuits of fast connectivity, and a fully externalized nervous system into the emergent infrastructure of the digital brain. Definitely not openly hegemonic and certainly not operating in the language of domination, the first encounter of neurobots and humans produces individuals who actually begin to see, think, and feel like the sociobots of their wildest dreams.

Psychic Robots

A BBC report titled "Robotic Prison Wardens to Patrol South Korean Prison" describes a prototype demonstration of prison guard robots that would monitor inmates for "risky behaviour," specifically suicidal tendencies and violent impulses.[10]

Quickly migrating beyond the use of robots to physically guard prisoners, this prototype project represents that moment when robots first began to evolve beyond their purely mechanical function as prison guards to the more complex task of carrying out psychiatric assessments of the behavioural patterns of prison inmates. While it could be expected that robots would first enter prisons in the traditional roles of surveillance and control, the three robots involved in the demonstration project have a very different task: namely, to mingle among a captive population as only a five-foot robot can do and, while "looking more friendly to the inmates," conduct an active search for signs of suicidal and violent behaviour. It's not so much, then, a demonstration concerning the feasibility of using robots in prison environments but actually an experiment with very general applications for perfecting an operating system allowing robots to conduct complex psychiatric examinations of prisoners. At this point, we move beyond cinematic images of prisons of the future with robotic guards in towers carefully monitoring prison populations to that moment when technology actively penetrates the human psyche in search of "risky behaviour." Here, robots are no longer mechanical devices, but artificial psychiatrists equipped with 3D vision, motion detection, and programmed operating systems, all aimed at discerning visible signs of melancholy, rage, despair, desperation, fatigue, and hopelessness.

While it is not evident from media reports how robots are to fulfil complex psychiatric examinations – other than the mention of the demonstration robots monitoring abrupt changes in the behaviour of individual prisoners – the intention is clear: for prison guard robots to cross the boundary between surveillance from the outside of captive bodies to internal explorations of psychic behaviour. Guided by a prescriptive doctrine concerning the parameters of "risky behaviour," what is really being tested here is robots as avatars of the new normal, conducting frequent visual examinations of a chosen, and necessarily captive, population in order to determine which bodies fall inside and outside of the normative intelligibility determined by the artificially defined ethics of "risky behaviour." In this case, it is the responsibility of those bodies placed under surveillance to provide no outward signs of either visible dissent (violence) or refusal of the state's power over life (suicide). While at first glance it might seem that guard robots are not programmed with levels of artificial intelligence and, perhaps, artificial affectivity necessary to detect otherwise invisible signs of powerful emotions internal to the psychic life of prisoners, what may be brought into *political* presence here is an entirely new conception concerning how power will operate in the robotic future: not so much the great referentials of power over death, or even power over life, but *power over visible expressions of human affectivity* – a form of robotic control that assumes the psyche is not a form of internal *being* but a kind of external *doing*; that is, the psyche is not something we have but something we do. In this scenario, what is important about the human psyche for purposes of the society of control is less the complexities of hidden intentions – the cultural acedia associated with feelings of melancholy, resentments that activate rage, total powerlessness that motivates despair – than those visible, outward manifestations of the rebellious psyche, that moment when the bodily psyche moves from the long, silent gestation of hidden intentionality to overt declarations of its intention to act, whether through violence or suicide. At that point, at least according to this prototype demonstration, robot guards will be waiting along the watchtower of the society of control, quickly targeting immanent signs of psychic rebellions against the order of normative intelligibility, relaying warnings to central command, all the while standing by for further instructions.

I, Robot Land (in Ruins)

This is a report about a spectacle in ruins, a "history of the future," which in its original imaginary form did not survive the future. In this case, the original proposal for the construction of Robot Land in Inchon, South Korea, involved the development of a vision of the robotic future that would seamlessly blend business, entertainment, and advanced robotic engineering – in effect, a spectacular showcase not only for Korean prowess in robotic engineering and innovative business practices but also for the global diffusion of the Korean imaginary throughout network society, including entertainment, design, and culture. That original history of the future did not actually survive the future. Indefinitely postponed with (tentative) plans now underway to undertake construction from 2022–24, the newest conception of Robot Land strips away most of its spectacular entertainment features, focusing instead on business and engineering interests. With this, fantasy collapses under the weight of reality; spectacle is liquidated by history; and the future is terminated by necessities of the history of the present. What follows, then, is not a description of Robot Land as something actually existing but as a beautiful speculative proposal – an imaginary vision of the history of the future, which in its aesthetic daring, in its fusion of business, art, and robotic engineering represents so well Korea's creative contribution to the history of the future of robotics. This is a report, then, of a history of the future – its dreamy vision, cultural entanglements, and historical origins, which was never actually realized but which, in hindsight, assumes an even greater significance as a story of futures won and lost, and, perhaps, that point where the Korean cultural imaginary parted ways with its own speculative imaginary. It's a report, therefore, of a history of the future that never existed historically but, probably for that very reason, endures as an imaginary monument to spectacles in ruins – Isaac Asimov's beautifully complicated, turbulent yet harmonious, vision of I, Robot *slams into the unforgiving reality of earthly material history.*[11]

In September 1950, Incheon, Korea, was the site of a daring, and justifiably famous, US invasion at the height of the Korean War, which aimed at capturing the capital city of Seoul and thereby decisively cutting off vital supply and communication lines to North Korean forces who were engaged in besieging United Nations forces further south in the Pusan peninsula. Identified as "Operation Chromite," as conceived by General Douglas McArthur and carried out by the 1st and 5th Marine Divisions, the invasion force successfully shifted the momentum of hostilities, eventually resulting in the present-day demarcations of North and South Korea.

Possibly as an unconscious tribute to the above invasion, Incheon was selected as the site of a second invasion, this time not by US Marines charging ashore but by astral landing craft carrying robots from the past, present, and future. The invasion force consisted of a multitude of creative robotic engineers, futurist designers, and marketing experts in entertainment spectacles, all aimed at successfully establishing a cutting-edge theme park called Robot Land, made up of robotics engineering displays, commercial applications, and futurist-oriented research facilities depicting the future of robot society, as well as possibilities for harmony among robots, humans, and nature.[12] Not so much a Disneyworld for robots, since that would entail focusing on a symbolically rich, but past-oriented, narrative of mass entertainment spectacles, Robot Land, as originally proposed, had a very different objective. Conceived as a "history of the future," the guiding ambition was to construct a theme park depicting a future robotic society that, while visually honouring the history of robotics engineering as well as visions of robotic society originating in science fiction and Hollywood cinema, would actively and very directly engage in the project of designing the robotic future. Here, the robotic future anticipated by business, engineering, cinema, comic books, and literature would be paralleled by state-of-the-art research facilities aimed at both confirming and promoting Korea's creative leadership in the areas of robotic design, fabrication, and engineering. Imagined as a gateway to the future rather than a spectacle of the past, Robot Land had chromite at its techno-visionary core, anticipating a hard-driving futurist invasion by the Korean robotic imaginary of global markets and perhaps of generalized cultural imagination as well. Part *theme park* (featuring a gigantic roller coaster that dangles off the arm of a gigantic robot before plunging into the water below; a robotic aquarium filled with robotic fish, including lobsters and jellyfish; and merry-go-rounds for riding robot animals), part *futurist robot laboratory*, and part *"industrial promotion*

facility," the original vision of Robot Land took seriously its mission to intensify the "fun and fantasy" in the robotic future. There were, of course,
necessary, indeed inevitable, exceptions, as in any story concerning the
unfolding (artificial) future. In the midst of this intended celebration of
robotic fantasy, there were also plans underway to demonstrate "how robots may be used in 2030, particularly when it comes to assisting seniors
with housework, medical check-ups and dementia prevention."[13] There
were also psychological and economic regional geo-national sensibili-
ties at play. In this case, no sooner had the Japanese constructed two
colossal robot statues (*Tetsujin* in Kobe and *Gundam* in Odaiba), than
Korea's Robot Land was conceived to trump Japan's claim to supremacy
in the area of gigantic robotic spectacles, with a strikingly colossal 364-
foot statue of Taekwon V (Voltar the Invincible).[14] In this case, persistent
and longstanding tensions between Korea and Japan found their most
recent manifestation in the twenty-first century in the delirious form of
robotic fiction.

Considered as "a history of the future," there was at least one significant, perhaps terminal challenge to the overall logic of the project,
hinted at by the very naming of the theme park – Robot Land. Possibly
conceived as a Korean alternative to the "magic kingdom" of California's
Disneyland Park, where "you can sail with pirates, explore exotic jungles,
meet fairy-tale princesses, dive under the ocean and rocket through the
stars – all in the same day,"[15] Robot Land, as originally conceived, offered
its own alternative vision of a future distinct from the Disneyland prescription with its "8 extravagantly themed lands – Main Street, U.S.A,
Tomorrowland, Fantasyland, Mickey's Toontown, Frontierland, Critter
Country, New Orleans Square and Adventureland."[16] While Disneyland
seduces by translating the phantasmatic ideology of the American dream
into nostalgic spectacles, Robot Land was meant to deliver a harder message: that robots are here to stay, whether taking the form of lobsters and
jellyfish, assuming the entertainment guise of robotic animals gathered
together for a fun carousel ride, inflating to the gigantic proportions of
apocalyptic cinema like the massive statue of Taekwon V, or, more prosa-
ically (but pervasively), spreading out their established robotic hardware
as the real working infrastructure of global automobile manufacturing
or, for that matter, as futurist technological prosthetics for the sick, the
aged, and the demented.

While Jean Baudrillard might once have noted that the seduction of
Disneyland is its convincing pretence that its fantastic simulations are
an escape from the real world rather than what Disneyland really is –
a perfect model of the real-time model of soft power, modulated violence, and crowd management – Robot Land was to be the technological

order after the age of simulacra. Here, there would be technologically enabled thrills – roller coasters dangling from the outstretched arms of massive robots – mesmerizing robotic spectacles, and spectacular feats of imagination, but no order of simulacra, no sense that the new order of robotics is anything than what it really is: a key component of the Korean version of the power of the new real. With its mixture of entertainment spectacles, industrial promotions, and a graduate school in robotics, Robot Land was to be a place where fun illusions and delirious spectacles would always be underwritten by a very visible undercurrent of dead-eyed economic seriousness of purpose and carefully orchestrated research visions of (certain) robotic futures. This notion of place is, of course, the proposed theme park's biggest problem: the future of robotics will probably have nothing to do with any territorial referent; certainly, it will not be a "land" in any physical or even symbolic meaning of the term but will most definitely constitute a new order of time – robotic time. In this case, Robot Time, rather than Robot Land, would probably be a more accurate description of the new epoch ushered in by all futurist robotic designs, from mass entertainment spectacles to the complex artificial sensors working the assembly lines of the manufacturing world. When the future of robotics, one already anticipated by contemporary developments, turns away from its ready-to-hand terrestrial manifestations – artificial fish, mega-statues, humanoid machines – and enters the databases of globalized networked culture as their indispensable artificial intelligence and machine-to-machine and machine-to-human communication, then we will recognize that we are not following a technological pathway that will lead to a certain place (Robot Land) but one that will lead towards a certain (robotic) order of database time that is networked, communicative, and neutral. As with all things having to do with theme parks, actually expressing such a fundamental eschatological rupture in the order of things – the displacement in importance of visible space by the invisibilities of (database) time – is challenging. Such a challenge is probably why, although it takes momentary refuge in the comfortable referential illusion of Robot Land, this proposed theme park is one that will probably always be known for the hauntological traces of its essential missing element – the once and future epoch of *Database Robot Time*. There are definitely no "magical kingdoms," no "fairy-tale princesses," no "pirates" – just a proposed theme park on the edge of the rising time of the East, which announces that, for all the psychic exuberance of its robotic fossils, from fish and statues to carousel animals, it is one Tomorrowland that will not be able to camouflage for much longer what is really taking place in this second invasion of Incheon: the newly emergent order of the time of the robots, with humans kept on standby as their necessary prosthetics. Now that the original vision of I, Robot Land has been discarded, we may never have an answer to the question concerning the time of robots, but we definitely know that, in the future as in the present, design based on speculative fiction is still at the mercy of real historical time with its rise and fall of both human and robotic imaginaries.

Database Robots

What happens when the evolutionary destiny of robots suddenly splits into two paths? One pathway continues that which has long been anticipated by scientific visionaries, cinematic scenarios, and science fiction – namely, the triumphant rise of a new robotic epoch invested with technological inevitability as successors to a putatively declining human species – while an alternative pathway materializes in which robots abruptly shed their mechanical skin, upgrade their artificial intelligence, and adopt the remote senses of network culture as their very own interface with the surrounding world of human flesh. What, then, is the future of robotics: sovereign technological automatons or database robots?

Projective thought focused on the first pathway has long been the subject matter of technological futurists. In his brilliant book *Mind Children*,[17] the technological futurist Hans Moravec establishes clear-cut timelines tracing the history of robots, from their first appearance as mechanical prosthetics servicing human needs to that quickly approaching singularity moment (approximately 2050) in which robots equipped with advanced artificial intelligence, articulated limbs, and full-array sensory data inputs are projected to become an autonomous species, not only thinking for themselves but, more importantly, making sovereign decisions concerning what needs to be done in the interests of the preservation of the (robotic) species. Anticipating that day of fatal reckoning in which robots, as the product of human imagination, just might be inclined to follow familiar (human) pathways of revenge-taking for the gift of (robotic) life that they can never pay back, Isaac Asimov, in his celebrated book *I, Robot*,[18] is ethically pre-emptive in anticipating a future race of fully autonomous robots that are invested, outside their conscious awareness, of the guiding moral edict, first and foremost, to do no harm to human beings. That Asimov's anticipatory ethics of robotic behaviour would be quickly shrugged off by robots exhibiting all the behavioural, emotional, and moral traits of their human progenitors is, of course, the privileged focus of the science fiction writer Bruce Sterling, who, in his cult classic *Crystal Express*,[19] eloquently and passionately scripts a future war of robots spanning many galaxies – a war in which a class of robots known as "shapers" and an opposing robotic tribe identified as "mechanists" engage in protracted combat where the key issues at stake are as profoundly ontological as they are fiercely political.

Culturally, we are already well aware of the history of the (robotic) future that will be traced by the first pathway. Like a form of generalized anticipatory consciousness, many years of cinematic history have provided dramatic images of the multiple permutations, internal and external, that will likely follow the sovereign regime of robotic logic. While most cinematic encounters between robots and humans are ultimately settled by spectacles of violent battles, a few actually hint at flows of symmetry, flows of harmony among robots, humans, and nature,[20] while the remainder often conclude with unsettled paradoxes, unfinished narratives, and promises only made to be broken. For example, the final, anguished speech by Batty, one of the pursued replicants in *Blade Runner*, powerfully and evocatively captures both the anguished human will to live and a courageous replicant's pride in star bursts that he has witnessed, distant planets explored, and inexpressible awe before the vastness of deep space. When Albert Camus first articulated the absurd sense in *The Rebel* as consisting of an all-too-human demand for meaning to which the universe answers with indifferent silence, he probably did not have in mind a future time in which the hunted-down replicants of *Blade Runner* would be commonly haunted by an existential sense of the robotic absurd, that moment in which genuine anguish by replicants over their programmed termination dates is met with the silence of nature's indifference. That we are already conscious of the blending of technological dynamism, real power struggles, and stubborn, complex ethical entanglements, which will probably constitute the material reality of the future life of the robotic mind, is explored everywhere in the history of cinema, including those classics of visual imagination such as *2001: A Space Odyssey*, *Alien*, *Metropolis*, *Westworld*, *The Day the Earth Stood Still*, *Star Wars*, *Star Trek*, *Robocop*, *Terminator 2: Judgment Day*, and, of course, that poignant narrative of human senescence and robotic ingenuity – *Wall-E*. Like a cinematically driven society eager to be haunted by its technological future, and certainly capable of quick ethical and political adaptations to the demands of the (robotic) day, we may have already war gamed the future, played and replayed it, spliced and remixed the fractures, bifurcations, and liminalities likely to follow the *Judgment Day* of the technological future. In this case, it is as if the first pathway to the future – the often-told story of human hubris and cyber power – has already taken place in our collective imagination, leaving us now to be fully absorbed in studying in advance the psychic entrails of that fateful collision of the human species with its emergent technological successor.

However, with robotic life, as much as with human life, only opposites are ever true. Consequently, if there can exist such a rich cinematic and literary vernacular surrounding the robotic future, that might be because that future may have already reached its furthest limit and already begun to move in reverse direction, not necessarily by way of a spectacular implosion but by a silent yet discernible shift in robotic intelligibility. Perhaps robots themselves have grown tired of their rehearsed cinematic portrayals, shifting direction away from the spectacle of powerful AI machines to the more prosaic, more pervasive, certainly more perverse, and genuinely more futurist enactment of the approaching world of database robots. That is what the opening stories in this narrative of robots trekking across the uncanny valley is all about: not so much a predictable future of human/robot deep-space encounters but a more complex story of database robots expressed variously as neurobots, psychic robots, and avatars of robot time. In this case, robots have already fully penetrated the human sensorium, from hijacking the process of automated labour to relentlessly hacking the senses.

Beyond visions of technological apocalypse featuring predatory struggles between space-bearing robots and instinctually driven humans, the migration of robots into the minutiae of social life has quickly evolved from multi-axis industrial robots – automatically controlled, multipurpose, and functionally reprogrammable – specializing in the automation of labour to swarms of cyberbots, fluid networks of AI agents privileging the automation of cognition. With a rapid increase in the world robot population (300,000 in 2000 to 2,600,000 in 2020),[21] industrial robots have swiftly been integrated into manufacturing processes, particularly those reprogrammable around automated labour that promises to deliver predictability and reliability, backed up by resilience, speed, and precision."[22] An increasingly technical future arises, therefore, in which the compulsory labour of armies of specialized robots quickly displaces labouring human subjects in many work processes: welding, shipbuilding, painting, construction, assembly, packaging, and palletizing. Here, the overall trajectory follows the traditional path of economic development, this time with robots beginning in low-skill, sometimes dangerous jobs that can be done automatically and remotely, and thereafter moving up the skill-set ladder of career achievement to assume high-skill, hypercognitive positions in network culture. That, at least, is the overall technological ambition, marred sometimes by disquieting reports such as what happened when robots went wild in a General Motors (GM) factory built on a field of robotic dreams, with the seemingly inevitable mayhem of technological utopia at hyper-speed: robots smashing windshields, painting each other, crashing into vehicles on the assembly line.[23]

While computer malfunctions in a manufacturing plant can sometimes be solved by simply pushing the reset button, what happens now
when computer glitches affecting the core system logic of the externalized nervous system take down key areas of social life, including banking,
health, identity, and warfare? Without sufficient evidence concerning
the consequences of the wholesale transfer of the human sensorium
to electronic databases controlled, for the most part, by machine-to-
machine (M2M) communication and endlessly circuited by data robots
serving as synapses of the ablated world of cognition, finance, medicine,
politics, and defence, contemporary technological society has quickly
rushed into outsourcing itself, literally parcelling out human identity
into data clouds, from digital storage of personal health information to
complex networks for circulating financial data. When entire computer
systems crash, sometimes as a result of overload stress and at other times
for reasons enigmatic even to systems engineers, the result is no longer
an unexpected disruption in assembly lines, but the sudden data eclipse
of core areas of externalized human cognition. When data go dark, it is
as if the body has suddenly been divested of its key senses – it is the jetti-
soning of externalized memory, the disappearance of electronic profiles
of the extruded financial self, the circulation of electronic information
concerning medically tracked subjects, or the substitution of recombinant, digital orifices of the eye, ear, taste, smell, and touch in the age of
the rapidly dematerialized body. While many cautionary notes have been
struck concerning the inevitable fallout from a future populated by the
fully ablated self, skinned with an externalized nervous system and possessing an order of (digital) intelligibility modelled after extruded consciousness, only now is it actually possible to measure the consequential
results of this basic rupture of human subjectivity. Lost in clouds of data,
communicatively overexposed, its identity outsourced by fast digital
algorithms, its autobiography uploaded by data streams always offshore
to the vicissitudes of individual experience, the real world of technology, particularly robotic technology, reveals that we may have made a
Faustian bargain with the will to technology. Whether through generalized cultural panic over the sheer speed of technological change or
perhaps an equally shared willingness to ride the whirlwind of a society
based on the literal evacuation of human subjectivity, we have committed
to a future of the split subject: one part a fatal remainder of effectively
powerless human senses and the other a digitally enabled universe of
substitute senses. In the most elemental meaning of the term, the technological future that spreads out from this fatal split of human subjectivity cannot fail to be profoundly and decidedly uncanny. While robots,
technically forearmed with indifference, coldness, and rationality, will

probably at some point and in some measure successfully trek across the uncanny valley, the human response to the growing presence of the (technological) uncanny in contemporary affairs is far less certain. For example, consider the following reports from the uncanny valley that is daily life in the shadow of robots.

Uncanny Bodies

A few years ago, there was a newspaper report that evocatively captured the feeling of the uncanny in the robotic future. Appropriately titled "SociBot: The 'Social Robot' That Knows How You Feel," the report focused on the underlying element of uncertainty that is often a sure and certain sign of the presence of the uncanny in human affairs. Described as a "social robot that can imitate your friends," the robot, one researcher noted, was "like having a real presence in the room," with static photos instantly transposed in living, breathing facial expressions and "eyes that follow you," all thanks to innovative software.[24] (Virtual) emotion finally stars in its very own technological platform.

In his classic essay "The Uncanny," written in 1919 and perhaps itself deeply symptomatic of the profound uncertainties that gripped European culture after the First World War, Sigmund Freud approached the question of the uncanny on the basis of an immediate refusal.[25] For Freud, the uncanny – *unheimlich* – does not denote a kind of fright associated with the "new and unfamiliar" but something else – still indeterminate, still multiple in its appearances, and illusive in its origins. Far from being "new and unfamiliar," the uncanny for Freud represented something more enduring in the human psyche, "something familiar and old-established in the mind and which has become alienated from it only through the process of repression" – namely, the continuing yet repressed presence of "animism, magic and sorcery" in the unfolding story of the psyche. For Freud, scenes that evoked the feeling of the uncanny were remarkably diverse: "dismembered limbs, a severed head, a hand cut off at the wrist, as in a fairy tale of Hauff's"; "feet which dance by themselves as in the book by Schaeffer"; the "story of 'The Sand-Man' in Hoffmann's *Nachstücken* with its tale of the 'Sand-Man who tears out children's eyes' and the doll Olympia who occupies an unstable boundary between a dead automaton and a living erotic subject"; the always enigmatic appearance of the double; the fear of being buried alive; and, of course, the constant fear of castration. For Freud, whatever the particular animus that evokes feelings of the uncanny, the origin remains the same – the return of that which has been repressed not only by prohibitions surrounding "animism, magic and sorcery" but also by

episodic fractures, unexpected breaks in the violence that human subjectivity does to itself to reduce to psychic invisibility the complexities of
sexuality and desire.

Now that we live almost one hundred years after Freud's initial interpretation of the origins of the uncanny, does the emergence of a new
robotic technology such as SociBots have anything to tell us about the
meaning of the uncanny in post-human culture? At first glance, SociBots
represent a psychic continuation of that which was alluded to by Freud –
a contemporary technological manifestation of the feared figure of the
double as "something familiar and old-established in the mind." For
Freud, what is truly uncanny about the figure of the double is not its
apparent meaning as mimesis but its dual signification as simultaneously being "an assurance of immortality" and an "uncanny harbinger
of death." That is, in fact, the essence of SociBots: an assurance of (digital) immortality, with its ability to transform a static photo into an animated face, complete with twitches, blushes, and possibly sighs; but also
a fateful harbinger of death, with its equally uncanny ability to transform
living human vision into what Paul Virilio once described as cold-eyed
"machine vision"[26] – machine-to-human communication with a perfectly
animated software face tracking its human interlocutors, twelve test subjects at a time. In this case, like all robots, SociBots certainly give off
tangible hints of immortality – upload a photo of yourself, a friend, an
acquaintance, and they are destined for eternal digital life. But as with
all visual representations come alive, it is also a possible harbinger of
death, provoking feelings of human dispensability, that the tangible human presence can also be quickly rendered fully precarious by its robotic
simulacra. Interestingly, while Freud began his story of the uncanny with
a reflection upon the psychic anxiety provoked by the figure of the Sand-
man, who robs children of their eyes, SociBots may well anticipate death
in another way, this time the death of human vision and its substitution
by a form of vivified robotic vision. Here, SociBots could be viewed
as providing, however unintentionally, perhaps the first preliminary
glimpse of the psychic theatre of the Sandman in a twenty-first century
digital device, but with an addition. SociBots resemble the myth of the
Sandman in a second important manner: not only, like the Sandman,
does this technology provoke enduring, though deeply subliminal, human anxieties over the death of vision, but it also draws into cultural
presence, once again, that strange figure of the doll Olympia with its subtle equivocations between dead automaton and living erotic subject. In
this case, the particular fascination of SociBots, with their almost magical
and certainly (technological) occult ability to animate "features down to
the subtle twitches and eyes that follow you around the room," does not

solely reside in its animation of death but in its manifestation of a world where objects come alive, with eyes that track you, lips that speak, and facial features that perfectly mimic their human progenitors. It is neither death by automaton nor life by the doll-like construction of SociBots but something else: this robotic technology is one that derives its sense of the uncanny by always occupying an unstable boundary between life and death, software animation and real-life visual conversations and tracking. In essence, the uncanniness of SociBots may have to do with the fact that they are a brilliant example of the blended objects – part simulacrum/ part database – which will increasingly come to occupy the post-human imagination. Curiously, while it might be tempting to limit the story of SociBots, like the mythic tale of the Sandman before it, to stories of the death of human vision or even to the fully ambivalent nature of blended objects, from dolls to robots, there is possibly something even more uncanny at play here. It might be recalled that Freud controversially concluded his interpretation of the uncanny with his own psychoanalytical insights concerning the *unheimlich* place as the uncanniness of "female genital organs": "This *unheimlich* place, however, is the entrance to the former *Heim* (home) of all human beings, to the place where each one of us lived once upon a time and in the beginning."[27] While making no prejudgment on the genital assignment of robotic technology, it might be said, however, that the story of SociBots has about it a haunting and perhaps truly uncanny sense of a premonition about a greater technological homecoming in which we are, perhaps unwittingly and unwillingly, fully involved. In this interpretation, could the origins of the SociBots uncanniness have to do with its suggestion that we are now in the presence of technologies representing, in their essence, possibilities for a second (digital) rebirth? The suggestion of the uncanny, therefore, is that SociBots may well inhere in their capacity to practically realize the once and future destiny of robots as born-again technologies.

Junk Robots in the Mojave Desert: Year 2040

What happens when no tech meets high tech deep in the desert of California?

Just up the road from Barstow, California, and far away from the crowds of Joshua Tree, there's a junkyard where robots go to die. It consists of one hundred or so cargo-sized steel containers packed tight with the decay of robotic remains. Everything is there: a once scary DARPA-era animal robot weighing in at 250 pounds looks forlorn bundled in a shroud of net; early cobots and autonomous robots can be seen huddled together in one of the containers waiting to be reimagined; broken-down industrial robots that have reached a point of total (mechanical) exhaustion from repetitive stress injuries abound; abandoned self-organizing drone hives are left to slowly disassemble in the desert air; swarms of mini-robots – butterflies, ants, and bees – lie discarded; mech/cyb(ernetic) corpses of robots made in the images of attack dogs, cheetahs, and pack animals are jettisoned; all finally untethered from reality and left to rust in the Mojave Desert. Most of the valuable sensors seem to be missing, but what remains is the skeleton of our robotic past. The only sound heard is the rustle of scattered papers drifting here and there with scribbled lines of startup algorithmic codes. The only visual is the striking contrast of the sharp-lined geometry of those steel compartments against the soft liquid flows of the desert, land, and sky. The overall aesthetic effect of this robot junkyard is a curious mixture of the desert sublime, with the spectral mountains in the background and dusty scrublands close to the watching eye, mixed with a lingering sense of technological desolation.

What's most interesting about this robot junkyard – interesting, that is, in addition to its lonely beauty as a tarnished symbol of (technical) dreams not realized and (robotic) hopes not achieved – is that it has quickly proven to be a magnetic force attractor for a growing compound of artists, writers, and disillusioned computer engineers. Like a GPS positional tag alert on full open, they come from seemingly everywhere. Certainly, they come from off-grid art communities on the plains of East Texas, some transiting from corporate startups in Silicon Valley, a few drifting in from San Francisco, probably attracted by the tangible scent of a new tech-culture scene; there are even reports of artists drifting in from around the global net – Korean robo-hackers, Japanese database sorcerers, Bulgarian anti-coders, and European networkers – taking up desert-style habitation rights in the midst of the robot junkyard. It's a place that some have nicknamed *RoVent* – a site where heaps of robots can be retrieved, repurposed, reimagined, and reinvented.

It is almost as if there is a bit of telepathy at play in this strange conjuration of the artistic imagination and robots in transit to rust. Instinctively breaking with the well-scripted trajectory of robotic engineers that have traditionally sought to make robots more and more human-like, these pioneers seem to prefer the exact opposite. Curiously, they commonly seem to want to release the spirit of the robots, junkyard or not, to find their own technological essence. What is the soul of a data hive? What is the spirit of an industrial drone? What is the essence of a junkyard robotic attack dog? What makes a beautiful – though now discarded – robotic butterfly such an evocative expression of vitalism? Strangely enough, it is as if something like a Japanese-inspired spirit of Shinto, where objects are held to possess animate qualities and vital spirits, has quietly descended on this robot junkyard with its detritus of technical waste and surplus of artistic imagination.

The results of this meeting of supposedly dead technology and quintessentially live artists are as inspiring as they are unexpected. For example, one artistic display consists simply of a quiet meditation space where some of the junkyard robots are gathered in a rough circle, similar to a traditional prayer circle or the spatial arrangement of an ancient dirge, all the better to find their inner *moe* or, at minimum, to reflect on that illusive point in their individual robotic work histories where the mechanical suddenly becomes the AI, the vital, the controlling intelligence and then, just as quickly, slips on backwards into the pre-mechanical order of the junkyard burial site.

In the darkness of the desert night, another artistic site is organized as a funeral pyre for dead robots. Without much in the way of wood around for stoking the flames, these artists have paid a nocturnal visit to the ruins of the CIA-funded Project Suntan, close to a super-secret aviation project, where a barrel of abandoned liquid hydrogen has been retrieved for releasing the night-time spirits of (robot) mourning. The funeral pyre should be a sombre place, but in reality it's not at all. Maybe it is simply the visual, and thus emotional, impact of a full-frenzy funeral pyre, fuelled by the remains of secret experiments in high-altitude aviation fuels, sparking up the desert air. Or then again, perhaps it is something different, something more decidedly liminal and definitely illusive. In this case, when robots are stacked on a burning funeral pyre, it is very much like ritualistic final consummation, that point where the visibly material melts down into the dreamy immaterial and even the scientifically contrived mechanical skins, electronic circuitry, and articulated limbs finally discover that their final destiny all along was an end-of-the-world return to the degree zero of flickering ashes. The concluding ceremony for this newly invented Ash Wednesday for dead robots is always the same: a meticulous search by the gathered artists for the final material remains of the robots, which are then just as carefully buried in the dirt from which they originally emerged. Ironically, in the liturgy of the funeral pyre, there is a final fulfilment of the utopian – though perhaps misguided – aspiration common, it seems, to many robotic engineers, namely a haunting repetition, in robotic form, of the human life cycle of birth, growth, and senescence.

However, if the stoked inferno of the funeral pyre for abandoned robots sometimes assumes the moral hues of an anthropomorphic version of (robotic) imagination, the same is most definitely not the case at a third site, where a feverish outburst of the artistic imagination – splicers, mixers, recombinants, recoders – plies its trade anew by remaking this treasure load of robot technologies. Here, strange new configurations emerge from creative remixes of self-organized drone hives and fluttering robot butterflies. When (dis)articulated robot pack animals, some missing a leg or two, are repaired with extra legs culled from leftover parts of robot dogs or now only two-legged robot cheetahs, the result is often spectacular. It is just as if, in this act of robotic reinvention, the drudge-like life trajectory of many robots, previously valued only for invulnerability to boredom, boredom with things (repetition) or boredom with human beings (routinization), is suddenly discarded. What's left is this genuinely fun scene of robots, forever heretofore consigned to compulsory labour, untethered from their AI leashes, finally free to be what they were never designed to be: robotic cheetahs moving at the speed of a just-reassembled pack animal; robotic attack dogs, now equipped with re-engineered robotic butterflies for better visual sensing, suddenly sidling away from high-testosterone attack mode in favour of startling, but ungainly, emulations of those exceedingly life-like Japanese theatrical robots. In this artistic scene, it is no longer the animating spirit of Shinto at work, but something else: the splice, the mix, the creative recombination of robotic parts into a menagerie of creative assemblages. Or maybe not. Some of the most fascinating projects involving this group of recombinant artists were those by descendants of *Survival Research Lab*. Their renderings quickly brushed aside the aesthetics of creative assemblages in favour of a kind of seductive violence, which is, it appears, autochthonous to the American imagination. In this scenario, it is all about riding the robots. Robots as monster dogs, cheetahs, wildcats, sleek panthers, and large-winged earthbound birds waging war against one another or at other times left untethered to roam the night-time desert, whether as sentries, mech watchdogs, or perhaps free-fire zone attack creatures burning with the ecstasy of random violence.

Designs for the Robotic Future

A Cheetah, an Android Actress, and the (AI) Cockroach

Intimations of the robotic future are often provided by the design of robots presently being assembled in engineering research labs in the United States, Japan, and the European Union. The robotic future is not visualized as fully predictable, determined, or, for that matter, capable of being understood as embodying an overall telic destiny but, much like the human condition before it, as something that will likely be contingent, multiple, and complex. Indeed, if robots of the future – presently being designed on the basis of advanced research in sensor technologies, articulated limbs, and artificial intelligence – provide a glimpse of that robotic future, then it may be that traditional patterns of human behaviour notable for their complex interplay of issues related to power, affectivity, and intelligence may be well on their way to recapitulation at the level of an emerging society of future robots. Consequently, while the ultimate destiny of the robotic future remains unclear, its possible trajectories can already be discerned in the very different objectives of remarkably creative robotic research. Building on traditional differences in approaches to technology in which the United States generally excels in software, Japan in hardware, and Europe in wetware (the soft interface among technology, culture, and consciousness), new advances in robotics design inform us, sometimes years in advance, concerning how robots of the future will effectively realize questions of (soft) power, (machine) affectivity, and (artificial) intelligence. For example, consider the following three examples of contemporary robotic designs, none of which fully discloses the future but all of which, taken together, may provide a preliminary glimpse of a newly emergent future in which human-robotic interactions will often turn on questions of power, emotion, and consciousness.

Robots of Power

In the cutting-edge research laboratories of Boston Dynamics, there are brilliant breakthroughs underway (mostly funded by DARPA) in designing robots that embody a tangible sense of power, robots with astonishing capabilities for moving quickly over a variety of unexpected terrains. For example, the *Cheetah* robot is described as "the fastest legged robot in the world," with "an articulated back that flexes back and forth on each step, increasing its stride and running speed, much like the animal does."[28] Its robotic successor, the *Wildcat*, has already been released from the tethers of *Cheetah*'s "off-board hydraulic device" and "boom-like device to keep it running in the center of the treadmill"[29] in order to explore potentially dangerous territories on behalf of the US Army. The *Cheetah* and the *Wildcat* are perfect robotic signs of forms of power likely to be ascendant in the twenty-first century: remotely controlled, fast, mobile, predatory. Google purchased Boston Dynamics[30] (possibly as a way of acquiring proprietary rights to its unique sensory software), which may indicate that important innovations in software development are themselves always sensitive to the question of power, seeking out, in this case, to ride Google into the robotic future, at least metaphorically, on the "articulated back" and fast legs of *Cheetah* and *Wildcat*.

Robots of Affectivity

In Japan, it's a very different robotic future. Here, unlike the will to power that seems to be so integral to the design of American versions of the (militarized) robotic future – whether terrestrially bound or space-roving robots like *Curiosity* on Mars – Japanese robots often privilege designs that establish emotional connection with humans. Japanese robots are the newest of all the "companion species." Focusing on robots specializing in therapeutic purposes (assisting autistic children, augmenting health care, helping the elderly cope with dementia) or for straightforward cultural consumption (androids as pop entertainment icons, robotic media newscasters), the aim has been to cross the uncanny valley in which humans begin to feel "creepy" in the presence of robots that are too human-like in their appearance and behaviour. Psychological barriers against crossing the supposed uncanny valley have not stopped one of Japan's foremost android designers, Professor Hiroshi Ishiguro, who, working in collaboration with Osaka University, has created a series of famous robots, including an android actress *Geminoid F*, described as "an ultra-realistic humanlike android" (who smiles, frowns, and talks), and, in a perfect act of simulational art, an android copy of himself.[31] While Boston Dynamics's *Cheetah* and *Wildcat* may provide a way of riding power into the future, *Geminoid F* and Professor Ishiguro's android simulacrum do precisely the opposite by making the meaning of robots fully proximate to the question of human identity itself. If there can be such fascination with android actresses and AI replicants, that is probably because Japan's version of the robotic future already anticipates a new future of robot-human affectivity, one in which questions of strangeness and the uncanny are rendered into indispensable dimensions of the new normal of the robotic future. In this sense, what is brought to the surface by the specifically Japanese realization of the full complexity of robot-human interactions is the very shape and direction of individual and cultural psychology in the future. While *Geminoid F*, the android actress, will probably never really challenge boundaries between the human and the robotic, since it only represents a direct extension of the theatre of simulation that is mass media today, an android replicant is something very different. When the alter ego finally receives physical embodiment in the form of an android replicant, the question may arise whether in fact an android "selfie" might potentially be perceived in the soon-to-be-realized robotic future as the very best self of all.

Robots of (Integrated) Intelligence

While American approaches to designing the robotic future often focus on questions involving the projection of power, and Japanese robotics research explores subtle psycho-technologies involved in robot-human interactions, European versions of the robotic future often privilege the complicated wetware interface involved when swarms of robots intrude into what, from a robot's perspective, are alien spaces, whether the industrialized workplace, human domestic dwellings, or animal, plant, and insect life. Much like the European Union itself, where the value of integration is the leading social ideal, EU-funded robotic research has quickly attained global leadership in its creative studies of the bifurcations, fractures, and complex fissures involved with the extrusion of robots into the alternative environments of humans, plants, objects, and animals. For example, European scientists developed a robot cockroach[32] "that smells and acts just like a roach, fooling the real insects into accepting it as one of their own."[33] The robot cockroach adapted their behaviour to living cockroaches, integrating themselves effortlessly into the larger cockroach community, even smelling like actual cockroaches thanks to the wonders of "pheromones" taken from insects and, once successfully integrated, acting like a modern-day robot pied piper by leading insects to "beams of light and congregating there."[34]

This research shows that the next stage of development for autonomous devices will involve constructing a blended reality of machines and animals for better cooperation in overcoming challenges and resolving problems, with machines and animals, in effect, sharing a common nervous system, highly attuned to the actions of one another – machines comprehending the animal universe and animals, in turn, sensitive to signals emitted by the universe of machines.[35] In other words, it's not so much a study of AI robot cockroaches drenched with pheromones (all the better to attract the attention of unsuspecting naturalized cockroaches) but a brilliant futurist probe into the new order of robotic communications, that point where robots learn to communicate with insects and, by extension, with plants, objects, and humans, and those very same plants, objects, and humans actually have a form of quick robotic evolution of their own by finally learning what it means to "perceive," "understand," and perhaps even "influence" the otherwise autonomous actions of robots. In this evolutionary scenario, a fundamental transformation in the order of communications, beginning with insects and then rapidly propagating through other species – plants, animals, and humans – anticipates a future in which the integrationist ideals of the European Union are inscribed, unconsciously, unintentionally, but certainly wholesale, on a newly emergent Robotic Union. The lowly cockroach, then, once coated with pheromones, is perhaps a fateful talisman of a possible future in which autonomous robots learn to "mimic" behaviour with such uncanny accuracy that humans, like those "darkness-loving creatures" before them, follow the multiple robotic bots of the future "towards the light and begin to congregate there." In this situation, the only question remaining has to do with the meaning and direction of the "light" of the robotic future towards which we are congregating, certainly in the future as much as now.

The Psycho-Ontology of Future Robots

The future of robotics remains unclear, still clouded by essentially transhuman visions projected onto the design of robots, still not willing or able to reveal its ultimate destiny, that point when robotic intelligibility takes command and in doing so finally begins to trace its own trajectories in the electronic sky. Yet, for all that, there is much to be learned from reflecting upon contemporary robotics design – lessons not only about robotic technology and creative engineering but also about that strange universe signified by complex encounters between robots and humans, which take place in otherwise relentlessly scientific labs around the world, from Japan and the United States to Europe. If the future will be robotic – at least in key sectors of the economy as well as network infrastructure – it is worth noting that the overall direction of that robotic trajectory already bears discernible traces of human presence, whether in terms of conflicting perspectives on robotic design or what might be prohibited, excluded, and disappeared from our successor species. Not really that long ago, an equally strange new phenomenon – the human "self" – was launched into history on the basis of key ontological conditions, some visible (the complex learning process associated with negotiating the human senses) and some invisible (the order of internalized psychic repression). In the same way, contemporary society witnesses, sometimes in mega-mechanical robotic expression and at other times in specifically neurological form, the technological launching of a robotic species that, while it may eventually possess its own unique phylogenetic and ontogenetic properties, will probably always bear the enduring sign of the human – not necessarily in any particularly prescriptive way but in the more enduring sense that the trajectory of the robotic future, hinted at in the creative designs of robotics engineering, may well culminate in investing future robots with a complex history of internally programmed psychic traumas that will powerfully shape their species identity, both visibly and invisibly. In this case, contemporary fascination with robots may have its origins in a more general human willingness, if not eagerness, to displace unresolved anxieties, unacknowledged traumas, and, perhaps, grief over the death of the human onto identified prosthetics, namely robots. Could the future of robotics represent, in the end, the ethical ablation of the human condition, including the sinister and the creative, the compassionate and the cruel, in purely prosthetic form? If that is true, are robots, like humans before them, born owing a gift – the gift of (artificial) life – that they can never repay? In that case, what is the future psycho-ontology of robots: unrelieved resentment directed against their human inventors for a gift of life organized around "compulsory servitude" or the supposed joy of (robotic) existence?

Epilogue: From the Insurgence of the Blended Mind to Gen Z

Data Flesh

If this era can be such a time of extremes – protests against police brutality, the resurgence of Aryan nationalism, bitter political divisions, religious fundamentalisms at war against secularism, LGBTQIA rights guaranteed in law in a few countries but savagely oppressed in many others, labour disciplined by primitive accumulation (work or starve) versus growing oligopolies of finance capital – perhaps it is because, in part, trajectories explored by new media have slammed into the human condition with such accelerating violence that the resulting social, political, economic, and environmental wreckage all around us is like so much free-floating debris after the blast. Key tendencies that were once only theorized as possible futures have now escaped the realm of imaginary futurism to become the alphabet of a society, indeed of a world, seemingly imploding. We see panic anxiety as the key psycho-ontology of contemporary times: spasms of seductive misinformation circulating everywhere through the interstices of information culture; human subjectivity effectively reduced to data trash; digital delirium as the governing mood of a technological society that advances with equal measures of magic and menace; massive digital platforms breaking the skin barrier like body invaders of the sci-fi future; and everywhere possessed individuals – possessed sometimes by reactionary beliefs, technological transcendentalism, or religious passions, but many literally possessed also by deep, pervasive, and mesmerizing flows of digital codes circulating through affect, imagination, and consciousness. In this drift culture, only the most extreme signals manage to break through the noise generated by flows of information, often taking the form of the hysterical male with his anger over the loss of traditional masculinist privilege but, at other times, rising white nationalism threatened by what is perceived to be a permanently threatening outside world and always, of course, breaking news about those psychic breakdowns in the digital stream of those who can't or won't keep up to the speed of the nervous breakthrough that is life in the wires. Overexposed, overcirculated, overinformed, over-ablated, the triumphant age of being digital with its dreams of universal connectivity has quickly made visible all the broken connections of the past – ethnic grievances, religious differences, unresolved political feuds, gender violence, sexual oppression, economic injustices, moral indifference to the lives of the asylum-seeker, the slum dweller, the immigrant – just as much as it has effectively stripped digital subjects of the saving vision necessary for navigating the future. When the dust finally clears from this gathering scene of the ruins, within and without, the only thing standing will likely be the will to technology with its prophetic talisman hard at work on the human remainder: the exteriorization of human consciousness, the generalized synchronization of emotion, the ecstasy of finally becoming object-like, and the virtualization of culture. In the gathering dusk, the codes of technology are like digital sunshine on a cold, grey rainy day.

This situation can be clearly seen in the turbulent events occurring in contemporary society with the very real fear over viral contagion mixed with the quick return of politics in the streets with protests against racialized violence, and all of it multiplied in its intensity many times over by the pressures of generalized economic recession, severe job losses, and coming financial defaults. It's literally a time of imminent social chaos, deep anxiety, palpable anger over racialized injustice, with the inevitable political backlash just waiting to express itself, and all the while what is absolutely strengthened by the crisis are technological platforms putting down the codes for the will to technology. Today, remote communication is the digital lifesaver for an educational system that has quickly chosen to disappear into Zoom; mobile communication with its rich array of digital devices provides an instant working infrastructure for working online and at a safe distance; the death of the face-to-face social rapidly gives way to streams of information, gaming, news flashes, and communications as the technical lifeblood of network society; and most definitely, the end of (traditional) work as we know it has been accelerated by the pandemic with its reduction of the labour force to "essential services," clearing the way for a future that will quickly link artificial intelligence, deep learning, and creative robotics as the coming labour force. Seemingly, everything moves now in the direction of intensified, functionally required technological platforms. Everywhere, of course, the present mood of political malaise, economic distress, and social isolation contrasts sharply with the rising gains of finance capital as the codes of capital accumulation move in precisely the opposite, specifically *virtual*, direction from social and political disturbances on the ground left behind. Literally, we are witness today to the profoundly historical moment when the will to technology takes off, quickly achieving escape velocity, dynamically and spectacularly, from the inertial drag pressures of the social wasteland. While it is simultaneously predator and parasite, magic and drudge, fast (codes) and slow (life), agent of creative destruction and equally visionary new designs for the future, no one can know with certainty what a future streaming the will to technology will look like. But again, there is really no need to be a traditional futurist in this (digital) case, since this is one time in which the future is already in our past. And that future – the future of fully exteriorized consciousness, synchronized emotion, the desire to escape subjectivity and become object-like, the triumph of virtual culture – is in its fullest measure deeply, immeasurably paradoxical.

For example, when the privacy of human consciousness has been broken wide open by the nutcracker of digital technology, the result is fully unpredictable in its incommensurability. While there may sometimes be an impulse for the mind to go off-grid by undertaking private pilgrimages away from the data stream – a generalized strike against the media – that is surely a challenging battle. After all, how do you rid consciousness of the codes of technology that are already embedded as ways of seeing, framing of perception, gateways to imagination, structures of reflection? More to the point, the contemporary situation is something in the way of split consciousness, which in practical terms implies pushing warnings about surveillance culture to the periphery of attention and thus maintaining at least the illusion of personal privacy while participating in life in the wires, whether reluctantly or enthusiastically and for multiple purposes. But in the end, how long can the stresses, tensions, and contradictions of split consciousness endure? As the philosopher George Grant once asked, how long can human beings endure the "plush patina of hectic subjectivity lived out in the iron maiden of an increasingly objectified world inhabited by increasingly objectifiable beings?"[1] Fully exposed, its every movement digitally archived, tracked, and recorded, split consciousness is a daily battleground of bubbling brains in a coded world. When the digital sensorium touches the highly sensitive matter that is consciousness, exposing and fully externalizing minds waking up to find themselves in the digital stream, everything is suddenly out in the open: Twitter wars erupting everywhere, nuances lost, competing ideas in constant circulation, images jammed together without any particular pattern. Here, consciousness escapes its previous abode in the cranial shell to become something liquid, flaring up, influenced by swirling streams of information, breaking news, broken data, riding the digital whirlwind like a raw exposed nerve, often overwhelmed but with nowhere safe to hide.

In the end, who can tolerate the deep wound of a fully externalized mind? Who can quickly recover from the sudden injury done to human consciousness moving at the speed of light? What happens to consciousness when it is suddenly transformed into a psychic playground for unedited reality: continuously exposed to the raw footage of terrible events without time for an intervening ethical mediation; YouTubing screams of anger, rage, and pain with no necessary understanding of their context; filtering unmediated images, advertisements, analysis, and all the fun of the media circus with the externalized mind's frame rate always stuck on wide open? What happens when the suddenly exteriorized mind becomes a data catcher, a mood drifter, a lurker in the digital stream, sometimes a vicious predator intervening actively in the scene, but usually a digital voyeur absorbing like an information hungry sponge all the spikes, ebbs, and unexpected swerves of life lived under the accelerating velocity of the data stream?

Perhaps as a response to the tensions of split consciousness, we are witness now to the first growing pains of the fully blended mind: externalized, ablated, exteriorized, fully exposed in all the TikToks, Instagrams, YouTubes, and Snapchats of the digital world but, for all that, immersed in streams of information actively worked over by the insurgencies of creative consciousness. Serving as its own counter-gradient to the streaming environment of technological platforms, blended consciousness forms communities of interest, sometimes wraps itself in all the delirium of the network, goes public with its likes but keeps its real thoughts to itself, perhaps coded as a follower but really never that at all, since the blended mind is something strikingly new in the social condition. Tentative at first, patterned by algorithms, streamed by the ruling platforms, the blended mind is always highly unpredictable in its moods, contingent in its purposes, and capricious in its loyalties. It is born a hacker of closed systems of surveillance, a data storm forcing technological platforms to move at speeds unimagined, a grand inquisitor often repurposing indifferent flows of information with demands for social justice, an uprising of counter-surveillance within the blockchains of power and capital. At the same time, the blended mind – consciousness entangled with the speed of the network – opens the data stream to all the dystopia of the contemporary human circumstance: racist tweets, political extremism, ethnic stereotyping, gender discrimination, sexual abuse. The blended mind is what happens when the full dimensions of the human experience surge through global networks of connectivity, instantly becoming core content for technologies of the new real. Simultaneously living in the shallows but also swimming in the deep, sometimes intermittent in its attention but, at other times, absolutely intense, maybe disengaged, indifferent, and disconnected but also, and often at the same moment, deeply engaged, committed, and fascinated by flashing points of connectivity, the blended mind is a historical experiment in progress: slow consciousness and fast data, complicated lives and sterile algorithms, a raw spectacle of the life instinct falling upwards into the data dump.

Perhaps it's no longer simply a digital universe of blended minds but something more generalized, namely blended bodies: emotions, subjectivity, imagination, perception living in a world of data made flesh. Like a technological makeover of the Christian Bible, which begins with the words "In the beginning was the Word" (John 1:1–3), the digital bible begins with the translation of the spoken word of divinity into the cold data of technological platforms. And, while the biblical story of the Word was recorded in prophecies of the Old Testament and gospels of the New Testament as the coming of the long-awaited messiah, for whose bodily sacrifice humanity would be indebted from birth to an eschatological debt that could never be repaid, the digital bible substitutes salvation by data, an epochal opportunity to shed the skin of the human in favour of the soft skin of the digital. When the story of this particular period in history is ultimately written, it will no doubt focus on salvation by running the numbers – data made flesh – as the animating impulse behind skinning the digital world with DIY bodies, surveillance that never sleeps, drone skies, robotic cyber-machines, singularity theory, artificial intelligence, deep learning, and biometric sensors everywhere. Here, digital devices are portals to the latest secular version of salvation, the instant connectivity of smart phones a re-enactment of a religious state of grace, artificial intelligence a gateway to transcendence, globally synchronized emotion a mass conversionary experience, desiring to become a data object drifting in the network a basic article of digital faith, and the fast streaming of virtual life as so many triumphant hosannas to the moment of digital redemption.

However, just like the story of Christianity before it, this newest re-enactment of the myth of *salvation* in digital form has its competing myths: one, the myth of *order*, is literally the attempt by power and capital to channel the explosive energies released by network connectivity into a new digital order maximizing political loyalty and economic profit; and the other, the myth of *freedom*, is network society pushed to its limits and beyond by a rising generation of blended minds, data flesh, and virtual perception, particularly by the generation of the young today moving at particle speed.

If it is as true now as ever before that parasites are always quick to the feast, that scavengers are often first to exploit the radically new, the vulnerable, the tentative first beginning, then the story of technologies of the new real is no exception to the historical rule. No sooner did the new salvation myth of *being digital* take hold in the twilight years of the twentieth century then it was immediately hijacked by standing powers, effectively streamed in the direction of technological platforms as a new digital ordering of the world. Politically panicked by the ability of fast-moving flows of data to instantly overturn the earthbound world of national sovereignty, landed citizenship, bounded economies, framed identities, and hardened borders, power everywhere responded to the challenge by issuing its own counter-challenge: constant, ubiquitous patterns of surveillance, new alliances between the national security state and major technological platforms, the gathering of biometric data taking the form of social credit in China and contact tracing in the West, always seeking to upload our digital shadow, patiently going through the data trash we leave behind for its telling hints about our secret motivations, hidden intentions, and questionable loyalties. Politics in the wires, then, became a scavenger hunt in the data stream in the interests of preserving the power of the new status quo. It is the very same with digital capitalism. Unlike the sovereign state, which is focused for purposes of its own survival on preserving sovereignty over fixed, time-bound territory at a historical moment in which real power has taken flight into the fluid, space-bound empires of the digital, digital capitalism instantly shed its basis in territorially bound bodies, manufacturing, and sales, striking out for the monetization of the power of the flows of network society. That's been the business history of the past few decades – the violent abandonment of blue-collar workers and their factory-based jobs in favour of the higher exchange value of offshore labour without a permanent home; transnational trade agreements making possible cheap, exploited labour abroad and high-market consumerism domestically linked together by the restless movement of robotic cargo ships; and the shedding of traditional manufacturing by putting down the digital hammer of business going online, either becoming transnational, wrapping itself in fluid flows of circulating capital, or not existing at all. That is the ascendant power of contemporary technological platforms – this new era of digital capitalism with its dynamic, here today/gone tomorrow face; its technological platforms that place their fiscal bets on bending the myth of digital salvation in the direction of killer apps, digital fantasies, beautiful imaginaries of new social media, the everyday flux of data made flesh. But having acquired unimaginable corporate wealth by servicing network society, linking technological platforms and willing subjects with digital devices, and thus enabling the global diffusion of digital reality, the economy is fated to ride the whirlwind of the data storm. In the present social crisis, the most striking expression of the new class reality of digital capitalism is the radical split between ascendant finance capital in global stock markets and very real economic distress in the streets, mass unemployment, degradation of social and health services, and that most anguished barometer of inner anxiety, despair, and hopelessness – the opioid crisis. Here, the winning classes of digital capitalism – the owners of technological platforms and their supporting technocratic class – take economic flight into a golden future; while the economic losers in the new digital order – blue-collar workers abandoned with the destruction of manufacturing; many white-collar workers soon to be discarded as accidental road kill on the way to a future of artificial intelligence and robotics; and the permanently dispossessed, that is, the poor, the asylum-seeker, the unwanted immigrant, the racially vulnerable, the ethnically disenfranchised, the gender outlaw – face devastation and a bleak future. No less aggressive than the national security state in its pursuit of data about its network subjects, digital capitalism actually propels itself forward by making intense, granular surveillance of the consumption habits of its digital subjects a very exploitable opportunity for massive capital accumulation in the form of relational advertising, with every Facebook post enabling capitalist exchange value, every Google search triggering a chain of capital accumulation, every social media activity an addition to a very marketable digital profile. While digital reality may have been inaugurated by all the conversionary enthusiasm, messianic commitment, and utopian data visions brought together under the mythic sign of digital salvation in the late twentieth century, this twenty-first century, spiked by the technological platforms of capitalism and power, has surely been marked by the eclipse of digital utopia with the swift, relentless channelling of that original enthusiasm for life in the wires as salvation into new forms of digital ordering, including the national security state and digital capitalism, which, if they fail to possess the original intensity and dreams of digital utopia, have the more prosaic, but seductive, quality of stabilizing the flow, isolating digital subjects within familiar borders including national identity and property, and channelling desire along all the programmed streams of power and capital. Here, the life-changing experience of being digital as a religious epiphany is successfully replaced by the pleasures, dreariness, and growing screen addiction of the new digital bourgeoisie with its disciplinary state, austerity economy, and virtual phantasmagoria. After digital utopia, we now have stasis; after the failed epiphany of digital salvation, we face the enduring reality of taking our place, quietly and without a murmur of dissent, in the new digital order.

So the story goes, until life begins again. That is the contemporary social crisis. Messianic visions of the singularity moment may still have very real momentum as the leading contemporary edge of the myth of digital salvation. The COVID-19 viral pandemic may have precipitated a great shakeout of the last vestiges of the pre-digital economy, accelerating tendencies already under way towards a new digital order typified by remote communication, automation of the service industry, a real world of artificial intelligence and deep learning, and all of it tightly controlled as the proprietary knowledge of technological platforms. Nonetheless, the stability of the new digital order and the seduction of life in the wires have been challenged by the insurgency that is life in the streets today, with its scenes of surging political protests from cities in the United States, Canada, and Europe to rebellions by Hong Kong activists against new security laws promulgated by the Chinese state. All the while, other conflicts, other insurgencies based on race, class, gender, ethnicity, and national difference, have broken out seemingly everywhere, effectively challenging the sovereignty of closed digital borders with demands – some fundamentalist, some visionary – for reimagining a future off-grid to powerful machineries of surveillance as well as to technocratic digital platforms. Power today might reside in effective control over the creation, programming, distribution, and policing of algorithmic codes, but still the sounds and sights of life are palpably pushing up from below, sometimes allowing sunshine to burst through cracks in the data shield or maybe bringing with them a bit of cold, driving rain from the outside, hot energy from the streets falling downwards, rising upwards, threatening to rupture the set-piece framework of the new digital order with all the contingency, indeterminacy, differences, and absolute solidarities and hostilities that the human, now suddenly post-human, condition can muster.

And not just that. There's also the upsurge of the enduring myth of freedom – left-wing, right-wing, or no fixed wing at all, fundamentalist or progressive, or maybe a bit of both – which has the objective power of speed on its side. For example, consider what the political theorist David Cook has to say about politics in the quantum world:

K-Pop moves in the unpredictability of the quantum world where individuals interact in such a way to defeat any closure. Here individuals act at high speeds in quantum space, and in doing so each exceeding the traditional concepts of culture, intelligence and control. K-Pop sets forth the truly creative interchange of performance to bring forth new cultures from the rising generation (and indeed even some of the older generation) defying any predictable analysis, defying any notion of performance limited just to the group bringing all into performance, defying any notion of Euclidean and political space upon which the surveillance state attempts to close down the individual. Quantum culture works at particle speed with action and communication at a global distance where the paths are moving in multi-dimensional space.[2]

AI inhabits the quantum world but, in a twist of the Turing test, it is the machine that asks questions to determine the human. The attempt to control the individual through data mining and surveillance fails. There's always quantum uncertainty in determining the time and place, and the nature of individuals. So K-pop is entangled in the culture but paradigmatic of action that escapes the control of the surveillance, turning quantum theory against quantum AI.

My sense is that the younger digital generation – Gen Z – is the first to truly grasp the full potential of quantum culture with its "particle speed," "paths moving in multidimensional space," and "action and communication at global distance." In the background of their instinctive insight that, in quantum culture, we are speed particles not flat-earth objects, uncertainty paths not data profiles, multidimensional (beings) across virtual space not classifiable surveillance subjects, you can just hear the sound of breaking digital glass as the closed world of "Euclidean and political space" is shattered by the upsurge of the quantum with the youngest of the young generation leading the way. If that is the case, as I believe it is, then all traditional bets are off. DIY bodies will soon be putting on the skin of the speed particle. For reasons of survival, surveillance systems will have to play catch-up with the game changer of people as paths of uncertainty in multidimensional space. Indeed, quite soon masters of the new digital order, trapped psychologically in maintaining the hardened borders of territorialized nation-states and the soft borders of deterritorialized empires of capital, may only have predatory drones, unblinking robots, and the never dozing codes of AI algorithms to depend upon in their last ditch defence of fixed "Euclidean and political space." Life in all its enigmas, desperation, and hope will have left them behind. Technology in all its inexorable power, energy, and beauty is going quantum, likely leaving the present order of digital things with its surveillance systems, predatory prying, and the stickiness of its relational advertising as so much dead media on the road glimpsed in the rearview camera. That is, of course, the hope inspiring the myth of digital freedom, a myth that grasps the blazing comet of speed particles moving across the sky of quantum culture as its fateful talisman. But, as with all things quantum, everything has its simultaneous opposite value; contradictions are the essence of life in the fast lane; and indeterminacy of the future with its shades of hopefulness and uncertainty is probably the most likely sign of the times. That is why, I believe, the upsurge that is quantum culture will play itself out as a coming collision between aspirations towards being multidimensional, indeterminate, and finally free and the harsh reality of the future as a dead memory machine.

Growing Up Quantum

In the digital age, there are many ways of growing up quantum, many ways of having a technological autobiography shaping the way we see, think, and feel in the fourfold of land, sky, earth, and water, whether in the workday streets of the city of data or in the permaculture of all the back-to-the-land movements, often off-grid, of the young generation, which are, I believe, hopeful signs of something really new beginning to stir in the digital imaginary. After all, the next digital generation, net savvy from a very early age and rising to prominence fast, have already travelled at escape velocity right through McLuhan's prophetic imagination to its other side, that intriguing side where McLuhan intimated that technology pushed to its extreme turns into pure magic. You can see this side everywhere in those new riders of the data storm with the worldwide popularity of young adult fiction: those great new mythologies of the *Maze Runner* and *Shadow Hunters*; those films like *Nerve*, where you submit your choice as watcher or participant and know, both ways, the pleasure of the virtual race and the cruelty of the anonymity of the net; and that new poetry of the young dystopian imagination, like Cassandra Clare's *City of Bones* and *City of Ashes*. The air we breathe might choke us with its digital dust; the background noise might consist of the waiting times for the singularity moment; but the young digital beats of today and tomorrow have already departed for another place. Their way of being in the digital world might move at the speed of all the media moods and texted images of Instagram and Snapchat, but it seems to me that they have already made an epochal break with the desert of Big Data in favour of a much more interesting world filled with werewolves, demons, and shadow hunters. It's just like McLuhan intimated, but with this difference: the seduction of dystopia not utopia. Intensify any technology to its extreme, and it will surely flip into its opposite. So then, we have a digital future that is going to have to bend if it wants to find a way of accommodating the deeply felt subjective reality of dystopia with its magic, sorcery, and demonology that is everywhere the sign of the digital times. Bending the curvature of the digital future towards the creative and critical imagination is the project of *Technologies of the New Real*.

Dead Memory Machine?

The stresses and paradoxes explored in *Technologies of the New Real* are only the latest expressions of a more enduring debate in the digital imagination. To reflect on technologies of the new real in terms of the messy complications of image noise versus the sterility of pure signal – the politics of the street versus the technological imperative – is only to return, once more, to a fundamental contestation concerning trajectories of the future explored in all its contradictions, riddles, power, and creativity by many digital visionaries, theorists, artists, designers, and practitioners. Now, working under the sign of quantum culture with its celebration of the "both/and," namely that simultaneous opposite changes of state – wave and particle – are the normal way of things, I conclude this book with a quantum turn of my own. Turning back from reflections on the future, I would like to recover a lost, but highly important, insight into the consequential meaning of technology from the past – not from a generalized past, but from the country, Canada, of my own autobiographical past; moreover, not from Canadian thought generally, but from a very particular, but now unfortunately forgotten and long not debated intellectual past, namely a critical insight about the future of technology generated by what is sometimes called the Toronto School of Communication. In so doing, I am mindful that the Toronto School brings together thinkers as diverse as Marshall McLuhan, Harold Innis, Eric Havelock, Northrop Frye, C.B. Macpherson, and Charles Norris Cochrane in sustained, but very distinctly independently conceived, reflections on the technological adventure that is the modern, postmodern, and now, suddenly, post-human culture.[3] I am also mindful that this trajectory of thought is clearly an intellectual product of its times, framing out of its concerns the immensely creative energies of Indigenous theories of technology, the urgent importance of Black perspectives on the maelstrom, and, for that matter, feminist contributions to the Canadian discourse on technology. With that understanding of its limitations in mind, I would still maintain that, sensitive to the larger ethical, philosophical, and political questions raised by the preceding essays – DIY Bodies; Power under Surveillance, Capitalism under Suspicion; Dreaming with Drones; and Robots Trekking across the Uncanny Valley – it is appropriate, indeed necessary, to connect this present account of the experiments, drives, and technological ambitions involved in digital futurism with pressingly urgent questions raised, in particular, by two leading representatives of the Toronto School: Marshall McLuhan and Eric Havelock. If we are quickly migrating towards a future that will be marked in part by digital reality as a dead memory machine, then there is no better guide to that future than this long-neglected debate between McLuhan and Havelock. Always a quantum thinker at heart, McLuhan would be ebullient at the coming to be of a quantum culture, multidimensional, indeterminate, and moving at particle speed. Against this, Havelock introduces an important cautionary note, namely that this future may also turn out to run at the speed of a culture trapped in dead memories. Consequently, to reawaken the debate between McLuhan and Havelock in the context of *Technologies of the New Real*, to position Havelock's *Preface to Plato* against the technological humanism of McLuhan's thought, is not simply to pit competing interpretations of oral versus literate against one another but, more decisively, to ask if, in the present context of technologies of the new real, the digital future does not, in fact, belong more to Havelock than McLuhan: a future of technological dystopia rather than the long-hoped-for utopia of technological desires. In this case, is the digital future most adequately captured by McLuhan's prophetic account of new, creative epiphanies of technological consciousness with the promise of a new universal community mediated by new technologies of electronic communication or by Havelock's more practical account of oral culture as a dead memory machine? Considering the implications for the future of society under the sign of technologies of the new real, it is my sense that what the Toronto School has to tell us, what it wrote and thought about before the advent of digital reality, contains an important, indeed decisive, clue to understanding the digital future – a debate of truly global significance.

The further and deeper the Toronto School of Communication travelled into classical history, the closer they approached the always enigmatic digital future: its fatal contradictions, spectral hauntings, and creative tensions. With intellectual perspectives that hovered between intimations of catastrophe and insurgencies of hope, McLuhan, Havelock, Innis, and Frye offer heightened perception concerning whether the digital future will be a brilliant blast of creative energy or a deeply conservative repeating machine of Big Data, fast algorithms, and artificial intelligence. If, in the end, McLuhan's prophecy of electronic culture as a fabulously utopian recovery of speech clashes directly with Havelock's chilling insight that oral culture has about it tangible traces of a dead memory machine, then that would mean we are presently living in the fatal fallout of a technological society theorized a generation in advance in all its tangible possibilities and broken dreams by the Toronto School.

If we are fast accelerating towards the cultural black hole opened up by the clashing perspectives, in particular, of McLuhan and Havelock, it would also mean that we should pay close attention to two attempts by the Toronto School to escape its crushing gravitational pressures: the exit of political economy, inaugurated by Harold Innis and C.B. Macpherson; and the exit of mythology, traced in all its complexity by Northrop Frye and Charles Norris Cochrane. So then, a critical and expanded viewpoint on the Toronto School of Communication is a creative pathway to understanding both the immensity of the cultural crisis implicit in their writings and the urgency of their later appeals to politics and religion as ways of overcoming the closed force field of the digital imaginary.

Hogtown

Since Toronto often prefers to view itself as *the* essential Canadian city, the city whose outsized political influence, consolidated media control, and dynamic economic power shape and reshape the larger destiny of Canada as a whole, this self-image would also imply that the Toronto School of Communication, particularly when expanded beyond McLuhan, Innis, and Frye to include the thought of Havelock, Macpherson, and Cochrane, would be the irreducible core of Canadian intelligentsia, the trajectory of thought that, in its fateful intimations, puzzling contradictions, and clashing perspectives, reflects the essence of the Canadian intellectual imaginary. Here, the very pressure of the city itself as a technological dynamo on the northern shores of the Great Lakes mixes with a classical cast of mind as familiar with the literary ruptures of James Joyce and the tragedies of Greek culture as it is with the great codes of the Bible to produce a formative series of critical insights concerning what happens when the complexities of the word meets the hygienic purity of code, when digital culture emerges to present itself as the newest iteration of the new "Finn cycle" of oral culture.

To begin on an autobiographical note, the hard-edged, hard-driving, and certainly hard-working image of Toronto as a contemporary Prometheus has, from a young age, deeply shaped my own intellectual imagination. As someone born and raised in Northern Ontario, with an outlook on life that, in its critical realism and perhaps unfounded optimism, unconsciously absorbed the ancient energies of the Laurentian Shield, my first and most enduring impressions of Toronto were fittingly enough set in stone by that radiating beacon of Canadian communication, the noon-hour news report of CBC radio, just when I would always arrive home for lunch from another morning at Nipigon-Red Rock District High School. This daily confirmation that all was well in the carefully coordinated operations of Innis's empire of time was followed by something more sanguinary, namely the daily price of hog bellies from the killing floors of Toronto slaughterhouses. At some deeply informing level, almost like a lived experience of the power of oral culture, I have always retained in my own imagination that more traditional name for Toronto, "Hogtown," that blast of Promethean consciousness on the shores of Lake Ontario with its equal measures of time and blood. So, despite all the feverish appeals to rethink Toronto under the sign of the creative class or all the powerful communicative symbolism generated by the web of multinational head offices, glassy skyscrapers, and market-intensive digital networking that is Toronto today, I still prefer to think of Toronto as Hogtown, a city that might have its official eye on playing a leading role in the empire of communication but which, like all other great cities before it, is still haunted by its contradictions. It is a city that has buried its rivers, suppressed its Indigenous origins, and turned the deep cold waters of Lake Ontario into air conditioning for downtown high-finance skyscrapers; a city with suburbs frightened by the sounds of history coming from the city streets and its downtown core perplexed by the politics of anger sparked by often reactionary middle-class suburban subjectivity; a city with its power elites very much trapped in anxiety structures of their own imaginative making – anxious about not measuring up culturally to the prevailing media standards of the universal homogenous empire of the United States and psychologically distanced from its situated history as a quintessentially Canadian city. It is a city that is economically pragmatic and politically moderate, with all the complicated characteristics of a genuinely remix economy, population, and temperament – the reality of Toronto as a Promethean city on the lake that works.

Streaming Culture with Homer and Plato as Pathways to the Future

Why is it that, in the midst of the techno blast of the Promethean city, the leading theorists of the Toronto School of Communication took their cues from events two thousand years in the making, looking to the origins of Greek culture for intimations about the technological future, always situating their respective perspectives on communication within the larger trajectories of theories of Western civilization? In a strange intellectual version of transubstantiation, they might have been physically located in the habitus of a university by the Great Lakes with its stock exchange and slaughterhouses, but their consciousness was fully present at the beginning of Western culture. Their theories of communication walked the streets with Homer, Plato, the Sumerians, and the Assyrians as their constant companions. Why? Probably because, not content with the normal paradigm-hugging science of contemporary theories of communication, thinkers like Havelock, McLuhan, Innis, and Cochrane realized just how deeply and strikingly the question of communication was linked to the larger historical destiny of modern, and now postmodern, society. Indeed, reversing the advice provided by Heidegger in *The Question Concerning Technology*, where he counselled that we can best understand technology not by looking furthest but by paying attention to that which is closest at hand, the leading figures of the Toronto School of Communication were able to provide such brilliant accounts of the communicative future because they were intellectual astrophysicists who travelled deeply into the blast of the big bang of Western culture, that decisive moment when the elemental codes of communication governing speech and writing suddenly rose to the surface of consciousness under the warring signs of Homer and Plato. Strangely enough, a brilliant generation of Toronto theorists of communication turned their attention away from the sounds and sights of the Promethean city that is Toronto today in favour of an immensely patient and erudite descent into the genealogical depths of Western culture, that moment when the mythic, inclusive, in-depth tribalism of oral culture was suddenly shattered by the cultural ascendancy of the written word, that instant when Homeric mythic imagination was eclipsed by the written text of Plato as the great code of Western culture. Now that we live in the shadow of what McLuhan liked to describe as the dramatic reversal of the Western mind, with the return of mythic, deep, relational, inclusive oral culture as the essence of the electronic medium, from television to all the data streams of social media, and with that the dramatic disruption of the literate imagination – new Homeric data warriors, not Platonic scribes – it is easy to assume that the story ends successfully in the here and now of electronic culture.

McLuhan as Counter-Gradient

McLuhan certainly thought the story ended there. An Irish Catholic, domiciled at St. Michael's College, whose Catholicism was of the universalist kind, attracted to James Joyce and the Vorticism of Wyndam Lewis, strikingly dissimilar to the Irish Protestants whose settlement in Toronto after their flight from the Great Famine in Ireland quickly transformed Toronto of the 1920s into what was known as the "Belfast of North America," McLuhan studied the cultural entrails of the deep-space collision of oral culture versus written culture, drawing the perhaps hopelessly utopian conclusion that, like all transitory eclipses, the bright sunshine of oral culture has come into its own again in the form of electronic media of communication. With this elemental conclusion in mind, he imagined into existence his own Homeric saga fit for the age of the electronic blast. Now that the electronic world wears the skin of McLuhan, we have become accustomed to the naturalization of McLuhan's insight as the elementary matter of contemporary technological culture: technology as the externalization of human consciousness, hybrid media, the ablation of the human nervous system in the form of a series of rapid amputations of the senses – sight, sound, touch, taste, smell – technology as a series of panic amputations all meant to preserve the autonomy of the human mind. Indeed, we are now only in the first tentative stages of realizing that McLuhan has become his own counter-gradient, that understanding digital culture requires reversing McLuhan's insights. In the age of artificial intelligence, DIY bodies, surveillance culture, and robotic swarms, it's no longer technology as the externalization of human consciousness but just the opposite – human subjectivity as formed now by the deep *internalization* of technological consciousness; not hybrid media outside ourselves, but DIY bodies as digital hybrids facialized by Facebook, connected by Twitter, with consciousness angled between Reddit and 4chan, with Grindr for desire – streamed, fluid, mobile, personality shards drifting in the data stream. It's definitely not a world of understanding media, but one of *mis-understanding media* as the key to an electronic culture prophesied in all its dynamism and astonishing complexities by McLuhan.

Havelock's Fatal Dissent on the Meaning of Oral Culture

If putting on the skin of McLuhan's *Understanding Media* is the world of communication today – an expanding digital universe of hybrid media, newly generated digital selves equipped with in-depth, mythic, relational consciousness, the stunning return of oral culture from the so-called "tyranny of print culture" – what happens when McLuhan's prophetic vision of reanimated oral culture is rubbed against Eric Havelock's *Preface to Plato*, a definitely more cautionary account of the deep conservatism surrounding the genesis of oral culture in its original Homeric version? For all the astounding premonitory power of McLuhan's thought, there remains the disturbing counter-conclusion reached by Eric Havelock based on his study of the same elementary texts of the Western mind. With this work, Havelock introduced a certain disturbance in the technological mind by providing a fatal insight whose inexorable truth rises again everywhere to challenge the complacency of the utopian digital imaginary.

Unlike McLuhan, Havelock was profoundly doubtful concerning the emancipatory qualities of oral culture. Where McLuhan visualized mythic consciousness as providing instant entry into the liquid flows of the data stream, Havelock provided a more pragmatic note of caution.

According to Havelock, oral culture does not challenge the fundamental rules of social order, and specifically, the strength of oral culture is to transform culture into a memory machine for passing, from one generation to another, the survival rules for the games of life, whether in the language of the ancient mariner or the postmodern data steersman. What was really mythic, in-depth, and inclusive about oral culture was not so much the medium but the pragmatic content of that medium – its carefully encoded instructions, whether taking the form of song or storytelling, concerning what was necessary for the preservation of the Greek, that is to say, the Western species. About this, Havelock is explicit. Linked as it is to the real language of power, to the language of the preservation of the species, oral culture is quintessentially conservative in nature. In Havelock's terms, oral culture is decidedly mimetic, didactic, and metastatic, not creative, a retrieval mechanism for preserving core social and cultural memories. That's the basic lesson of Havelock's influential treatise, *Preface to Plato*. It is my sense that what Havelock has to say about the fundamental clash of perspectives between Plato and Homer, between the death of oral culture and the rise of an entirely creative surge of solitary thought that was the age of writing, is not something lying long in the past of digital culture *but in its future*. If McLuhan was correct that the electronic age unleashes a new Finn cycle of oral culture – tribal, mythic, relational, in-depth – it is also true that this newest version of oral culture is driven by a logic that is mimetic, indexed, preservationist, didactic, and repetitious – the data storm as an aggressive memory machine. That's the hyper-surveillance world theorized best by Edward Snowden: Big Data as the emblematic sign of didactic culture; the vast data archives through which our digital selves are processed as dead memory machines; deep learning machines as indexical, mimetic, and didactic; and everywhere the processed world of data as an aggressive memory machine drilling down deep into the granular matter of human subjectivity. McLuhan's thought in reverse image might be the *form* of the future, but the *content* of that future has already been foreshadowed in all its mimesis and didacticism by the brilliance of Havelock's account of the big bang of Western culture that was the meeting of the clashing worlds of Homer and Plato. The digital future is really about the breaking out again of a fatal split in the Western mind, this time with the Homers of the digital wilderness taking us home to the truth of digital being as a society of dead memories – electronically reanimated, remixed, recirculated but, for all that, a data storm that is cold to the degree zero and powerfully seductive because, in continuing the fatal split that is the essence of the great code of Western culture, it disrupts nothing. Welcome to the digital hospice.

The Data Archive as the New Cultural Storybook

What then is the digital future? Is it a transformative singularity event sparked by the fusion of human and machine subjectivity or our entrapment in a deeply coded culture – a new cultural storybook – that is necessarily didactic, preservationist, and mimetic in character? We know that McLuhan was alert to this split. He warned that tribal culture is also a closed culture, but for all that, he placed his future bets on the creative power of human imagination, our ability to serve as brilliant counter-gradients to the closed world of power. For all the apparent differences between Havelock and McLuhan, one a cultural genealogist and the other a technological futurist, there remains a lot of Eric Havelock in the thought of Marshall McLuhan. While Havelock argued that the fundamental division of Western culture, that between the age of poetic speech and necessarily abstracted writing, mirrored a larger divide between a Homeric state of mind and a Platonic world hypothesis, McLuhan did him one better. He actually provided a solution to the clashing worlds of speech and writing by making of himself a freshly minted Plato for the emerging electronic future of our new Homeric times. Everything in McLuhan's thought deeply grasped the truth of Havelock's essential insights concerning the profoundly conservative quality of oral culture: its closed tribalism; its disappearance of integral personality in favour of the group mind; its psychic surgery, then and now, on the human nervous system; its strange quality of mixing epics, sagas, stories of vengeful gods, the beautiful visions of fantasy literature with the ruling narratives of the great codes. A product of print culture, a bookworm living in the howling wasteland of digital delirium, a detached observer with highly attuned critical intelligence and reflective consciousness riding the data storm, McLuhan wasn't just the latest expression of a Platonic state of mind but something different, namely, a very unique fusion of Homer and Plato, a purveyor of the written word, making the hard stuff of words bend, split, and tense up under the pressure of his insistent questions, a poet of the electronic noosphere, an authentic personality who adamantly refused the closures of the new tribalism by making of his writing, thought, and art a powerful counter-blast against the cultural amnesia, political nostalgia, and numbed consciousness presaging the gathering digital maelstrom.

Now that we live in the technological dust of the insights of McLuhan and Havelock, necessary lessons can be drawn. The new world of ubiquitous surveillance, micro-granular data tracking, monitoring of the history of individual experience as so many shipping containers moving across the deep oceans of codes, and the construction of DIY bodies fit for fast travel through the data stream all point to the future intimated by Havelock, with McLuhan as his critical interlocutor, grasping deeply into his thought the emerging power of the data archive as the new cultural storybook but refusing to give up on the saving power of the artistic imagination. Of course, the fatal flaw in McLuhan's perspective was that the very moment of decisive cultural transformation, which he viewed as so creatively disruptive, the reversal from written to oral culture under the sign of the electric storm, was in Havelock's version of events precisely the founding act of a new wave of cultural domination. Could it be that it is precisely the creative energies released by the data storm – its hybridity, liminality, boundary shattering events, its great disruptors – that now serve most ironically as the tools by which the new cultural storybook of the data archive enters most deeply into the recesses of digital subjectivity? If so, McLuhan's lasting legacy would be the ironic one of facilitating the form of utopian technological consciousness necessary for realizing the predatory power of the data storm. In this case, technological humanism is the affect necessary for fusing consciousness with a digital storybook that is, like its ancient predecessor, mimetic, preservationist, didactic, and imitative. Again, is this possibility perhaps the reason why two of the leading contemporary philosophers of technology, Baudrillard and Virilio, have issued urgent warnings about the growing darkness of the digital future? Consider Baudrillard's notion that the radiating power driving the growth of technology is the spectre of a double death – the death of the social and the death of politics – and think about Virilio's envisioning of technology as inducing a generalized accident that takes down with it the overall metaphysical framework of time and space itself. While these critical perspectives might be melancholic outriders in a contemporary scene that witnesses seemingly unimpeded enthusiasm for the universal and homogenous technological state, their insights might also work to confirm the essential truth of Havelock's claim that the real meaning of the culture of connectivity is the deep connection of human desire, imagination, and activity, with technology as a dead memory machine, a new cultural storybook that can be so fascinating precisely because of the speed of its detrital remainder.

Digital Disorder

That Havelock is counter-gradient to McLuhan, with his vision of information culture as a dead memory machine counterposed to McLuhan's hopeful vision of universal epiphanies of technological consciousness, is confirmed by the trajectory of thought in the Toronto School of Communication that came after its foundational debate between Havelock and McLuhan. Here, Harold Innis's attempt to rehabilitate the question of technology with his "A Plea for Time" is undermined by the growing realization that the entanglement of the great referentials of technology, power, and capitalism in the contemporary era have produced a colonially hegemonic, territorially focused, space-binding Canadian society, bound together with the social glue of C.B. Macpherson's "possessive individualism." The defeat of poetic consciousness is indicated by Northrop Frye's abandonment of the "fearful symmetries" of Blake and his turn to the great code of the Bible. That religion is less a great code than a directly experienced resolution to a growing metaphysical crisis of Western culture is borne out by the insights of Charles Norris Cochrane, who, in his classic text *Christianity and Classical Culture*, argues that the Christian resolution to the great crisis of Western culture – the split between the spirit world and material reality, between passion and reason – is now breaking apart. In this sense, Cochrane jailbreaks Frye's vision of the great code, revealing that the code runs on empty, that the fatal scission of Western culture between passion and reason has erupted once again as the once and certain digital future.

Everything in the Canadian imaginary is fully present in the writings of the key intellectual figures of the Toronto School of Communication. While the Toronto School differs sharply from the Frankfurt School, particularly the critical theorization that is Adorno and Horkheimer's *Dialectic of Enlightenment*, in taking the grammatical turn from speech to writing as the foundational trope of Western culture, and while it was never really content with the technological closure of the American mind that is so enthusiastically expressed in the ideology of technological singularity, nevertheless there's much of America and Europe in the Toronto School. It is *futurist* because it represents a generation of thinkers who are quintessentially Canadian in their cultural awareness of the dynamic power of technological empire, and it is *genealogical* because of its desperate search in the archives of classical Greek culture for interpretative keys to the contemporary technological predicament. Ultimately, however, the Toronto School breaks with both, because it introduces a fatal doubt about the future of technology under the sign of the spirit of instrumental activism that is the pulsating core of the American mind, and rather than succumb to the European pull towards the particularity of historical consciousness, it prefers to live and think at the edge of the violent event scene that is the technological phantasmagoria.

What's the situation now that we live in the technological world envisioned in all its dangers and possibilities by the Toronto School of Communication? Quite literally, the "global village" is our newly naturalized habitus; the contemporary media scene is an acoustic bedlam of sometimes bitterly clashing rhetoric; the cultural storybook of digital reality, with its obsession with connectivity, distributive consciousness, technologically accelerated bodies spinning around in random whirlpools, and static eddies of information, is part of the new governing form of cultural intelligibility; and even digital subjectivity has got into the act by learning to live as an algorithm, to speak like a data stream, to feel just at the edge of terminal numbness and buoyant team-leader enthusiasm – the digital self as a startup always running on empty. Fittingly, like the Homeric state of mind that is *The Iliad* and *The Odyssey*, there are a lot of unexpected and definitely intractable problems in the global village. The blast of technology has wiped away print culture to reveal a strange new world of deeply intimate and just as deeply antagonistic clashes among ethnic rivalries, religious prejudices, unsettled political grievances, and atavistic feuds. Post-McLuhan, what we actually hear with our ears is a long repressed world of speech coming to life, from the fierce anger of ISIS and dreams of restoring the political form of the caliphate and neoliberal military remakes of medieval crusades to unpredictable spasms of nostalgia, anger, malice, and growing indifference. It doesn't help that Innis's "bias of communication," with its prediction of an ongoing political clash between technologies of space and time, has quickly become the ruling formula of twenty-first century political experience. After all, what is the meaning of newly resurgent English nationalism with its Brexit-framed refusal of the EU than a time-based rebellion of cultural and economic preservationists against the space-binding extensions of the European Union? What is the ongoing battle between the expanding spatial empire of NATO and the time-based Eurasian ideology that is the essence of Russian nationalism than the most recent example of Innis's eloquent account of the bias of communication in *Empire and Communications*? Everywhere, political elites scramble to take their place in McLuhan's prophecies, with some leaders like Justin Trudeau comfortably inhabiting the skin of coolness – in-depth, participatory, putting on the mask of the Canadian audience, definitely not playing a role but actually being physically, in appearance, clothing, and language, the *change* that seemingly everyone wanted to happen. Of course, in the cold political games of hot and cool politicians, coolness is also the natural media quality of Donald J. Trump, the seductive, quasi-fascist trance that attracts certain fascination by the media as they seek to fill in what he has left vague, to provide him with the definition that he otherwise avoids. Like a cool chameleon for a time of heated politics, Trump has that other quality that McLuhan considered so indispensable. He has made of his media personality a brilliant example of sudden reversibility, abrupt changes, an ongoing Twitter rupture with the status quo of which he was himself, as US president, the emblematic establishment representative, absorbing into himself all the contradictions, controversies, and always much sought-after scandals of the media stream. All the while, the many mass societies occupying media space today have their attention spans tracked, probed, recorded, and prodded as part of a polling culture that itself always acts in reverse, with the answer coming before the question.

Blue-Collar Blues for a Bootleg McLuhan

The Toronto School of Communication is the Canadian imaginary in all its intellectual fullness and spectres of desolation, both in what it privileges but also in what it excludes from attention. Now I know this to be true, first and foremost autobiographically, having grown up 700 miles north of Toronto in Red Rock, Ontario, a pulp and paper mill town on the northern tip of Lake Superior with the northern lights for a sky, the hard granite of the Laurentian Shield as a ground – the hard rock and equally hard-working firestorm of a single-industry, staples commodity kind of town for a soon-to-be-disappearing community – and, of course, with the cold blue waters of the deep, deep lake just about everywhere. Long before I actually read McLuhan, Innis, and Havelock, I grew up Canadian in the lived materiality of their thought. For example, my real instruction in *Understanding Media* was not theoretical but a matter of the blue-collar blues: a summer job working in the pulp room at the paper mill, reading a bootleg copy of McLuhan at break time but the rest of the time just doing my job, carrying a pick and cleaning bark off hundred-foot spinning drums of razor blades for skinning the logs fresh from the lake, trying not to fall into the drums, which was the fate of my predecessor who to this day is called "One Turn, Turner" because he fell into the drums and survived to tell the tale, popping salt tablets to fight the heat, and all the while, saying to myself: "The Medium is the Message" – what does McLuhan mean by that? Same for Innis: what he wrote about single-industry towns opening and closing at the speed of a capitalist economy organized to the rhythms of staple commodities was a chillingly accurate biography of all the Red Rocks of Canada, towns that quickly rise and disappear at the speed of extraction of value from all the staples of the world. In my experience, it's really only if your life fate has been closely tied to the precariousness of a staples economy and everything that that entails culturally and socially do you really grasp Frye's *The Educated Imagination* as a manifesto for another way of creative being. After all, which Canadian does not have a little bit of Eric Havelock inside their thinking – that nagging sense that the much-vaunted oral culture of all the great religions and mythologies and, most recently, digital culture is just a mnemonic, a repeating machine for learning the necessary lessons of power and discipline with severe punishments waiting for those who forget? *Technologies of the New Real*, then, is both a critical probe of the digital future, with its surveillance machines, drone warfare, DIY bodies, and robot drift, but also a way of taking seriously the ethical turn that is the deepest impulse of the Canadian algorithm and extending the lessons of the latter for a more critically engaged understanding of the world today. In the twenty-first century that lies ahead with its likely extinction events, my sense is that there is a desperate requirement for a form of thought that reflects this ethical turn, that travels deeply into technologies of the new real at the creative intersection, *the indispensable in-between*, of McLuhan's hopefulness and Havelock's dissent.

Formulated on the northern margins of American empire and long neglected in the noise surrounding the coming to be of the triumph of digital reality, the debate between McLuhan's utopian vison of the future as a possible "global village" and Havelock's more foreboding sense of the triumph of technological society as a "dead memory machine" is, I believe, the basic code war, the outcome of which will determine the shape of the twenty-first century. Once successfully extracted from the archives of the Canadian intellectual imagination and generalized across digital reality as a fateful talisman of the future, this debate is really what is at stake in the contemporary uprising that is Gen Z and the triumph of the blended mind. Here, technological society as a dead memory machine – a society stuck in traditional patterns of racism, inequality, discrimination, and destruction of nature – is beginning to be challenged by a generation coming to maturity with superb digital skills matched by an equally evocative sense of social injustice. Will the future bend in the direction of technological platforms with their vested interest in the digital status quo, namely a dead memory machine based on maximizing the accumulation of power, profits, and influence? Or will the future be sparked by a rising generation of teens and pre-teens, Gen Z, who, representing as they have from their digital birth in social media the very first real long-term inhabitants of the global village, will embrace a radically different vision of data made flesh, namely a technological society of the future, hopefully the near future, which begins to add the flesh of anti-racism, the skin of climate change, the passions of class struggle, the awareness of the meaning of deprival to the otherwise cold, sterile world of data. That Gen Z is a harbinger of the quantum generations of the future, with its ability to move very fast, to effortlessly negotiate the multidimensional worlds of social media, to ride the waves of digital contagion with the speed of a particle, and at the same time to be relational, networked, deeply entangled in all the complexities and complications of life online and off-grid, is as indisputable as it is massively hopeful. Of course, if the unfolding world of technologies of the new real is itself an unconscious retelling of a much more ancient struggle among the myths of salvation, order, and freedom, then the ultimate destiny of the digital future will itself be deeply entangled in the rise and fall of those classic myths across the wide-open, turbulent, fully unpredictable vectors of technologies of the new real.

Notes

Introduction: Viral Contagion and Death of the Social

1 Craig Fahner, "What Is Dystopia, Really?" *-empyre- soft-skinned space*, 16 May 2020, 22.32, http://lists.cofa.unsw.edu.au/pipermail/empyre/2020-May /011162.html.

2 Franz Fanon, *Black Skin, White Masks* (New York: Grove Press, 1967); Glen Sean Coulthard, *Red Skin, White Masks* (Minneapolis: University of Minnesota Press, 2014); Hamid Dabashi, *Brown Skin, White Masks* (New York: London: Pluto Press, 2011).

3 Gilles Deleuze and Felix Guattari, *A Thousand Plateaus: Capitalism and Schizophrenia*, trans. and foreword Brian Massumi (Minneapolis: University of Minnesota Press, 1987), 229–31.

4 Fahner, "What is Dystopia, Really?"

5 Hannah Arendt, *The Origins of Totalitarianism* (New York: Harcourt, 1967).

6 Vortis, "God Won't Bless America" (Thick Records, 2007), https://youtu.be /Dh9TqfYb-Yo.

7 For a brilliant description of Vorticism, see, particularly, the band's website, https://www.angelfire.com/indie/vortis/.

8 Vortis, "Vortis Manifesto," https://www.angelfire.com/indie/vortis /manifest.html.

9 Ibid.

10 Ibid.

11 Mitchel Anderson, "Humans Need a Prime Directive, Fast!" *The Tyee*, 7 December 2018, https://thetyee.ca/Opinion/2018/12/07/Humans -Need-Prime-Directive-Fast/.

12 See, particularly, Paul Virilio, *The Information Bomb* (London: Verso, 2006); and Marshall McLuhan, *Understanding Media: Extensions of Man* (Toronto: McGraw-Hill, 1964).

13 Quoted in Virginia Smart and Tyana Grundig, "We're Designing Minds," *CBC News*, 3 November 2017, http://www.cbc.ca/news/technology /marketplace-phones-1.4384876.

14 For a comprehensive account of ablated consciousness as part of McLuhan's description of the trajectory of technologies of electronic communication, see McLuhan, *Understanding Media.*

15 For an eloquent account of the cultural fallout of "duplex consciousness," see Paul Virilio, *L'horizon négatif: Essai de dromoscopie* (Paris: Editions Galilée, 1984). Jean Baudrillard's theory of simulation is traced in all its depth and complexity in Jean Baudrillard, *Simulations*, trans. Phil Beitchman, Paul Foss, and Paul Patton (New York: Semiotext(e), 1983). Michael Weinstein's interpretation of the externalization of mind is developed in depth in Michael A. Weinstein and Timothy M. Yetman, *Mind Unmasked: A Political Phenomenology of Consciousness* (New York: Routledge, 2017).

16 Paul Virilio, *Pure War* (New York: Semiotext(e), 1983), 44.

17 David Sanger, Julian E. Barnes, Raymond Zhong, and Marc Santora, "In 5G Race with China, US Pushes Allies to Fight Huawei," *New York Times*, 26 January 2019, https://www.nytimes.com/2019/01/26/us/politics /huawei-china-us-5g-technology.html.

18 Ibid.

19 Franz Neumann, *The Democratic and Authoritarian State: Essays in Political and Legal Theory*, edited by Herbert Marcuse (Glencoe, IL: Free Press, 1957); and Franz Neumann, "Anxiety and Politics," *Triple C: Communication, Capitalism and Critique* 15, no. 2 (2017): 612–36, https://doi.org/10.31269 /triplec.v15i2.901.

20 For a desolate account of cultural fatigue as the necessary precondition for the appearance of the "maggot man," see Friedrich Nietzsche, *On the Genealogy of Morals*, translated by Walter Kaufmann (New York: Vintage Books, 1989), 43.

21 Perry Sprawls, "Image Noise," in *The Physical Principles of Medical Imaging*, 2nd ed., web-based version, by Perry Sprawls, http://www.sprawls.org /ppmi2/NOISE/#INTRODUCTION%20AND20OVERVIEW.

1. DIY Bodies

1 Vernor Vinge, "The Coming Technological Singularity: How to Survive in the Post-Human Era," *Whole Earth Review* (Winter 1993): 11; Ray Kurzweil, *The Singularity Is Near: When Humans Transcend Biology* (New York: Viking Books, 2005); Kevin Kelly, *Out of Control: The New Biology of Machines, Social Systems, and the Economic World* (New York: Basic Books, 1992).

2 Francis Bacon, *Novum Organum*, edited by Thomas Fowler (Oxford: Clarendon Press, 1878); Ray Kurzweil, *The Singularity Is Near: When Humans Transcend Biology* (New York: Viking Books, 2005).

3 Raymond Kurzweil, "Reinventing Humanity: The Future of Machine-Human Intelligence," *The Futurist*, March–April 2006, 39–40, 42–6, http://www.singularity.com/KurzweilFuturist.pdf.

4 Martin Heidegger, *The Question Concerning Technology and Other Essays* (New Haven, CT: Yale University Press, 1977).

5 For a full elaboration of the breaking down of traditional divisions between the biological and mechanical, see Kurzweil, "Reinventing Humanity," 40.

6 Kurzweil, "Reinventing Humanity."

7 Ibid.

8 Gary Wolf, "The Data-Driven Life," *The New York Times*, 2 May 2010, https://www.nytimes.com/2010/05/02/magazine/02self-measurement-t.html.

9 Paolo Saraceno and Renato Orfei, "From Molecular Clouds to Stars," *Istituto Di Fiscia Dello Spaizo Interplanaterio*, CNR, Proceedings of "The Bridge between the Bang and Biology," Stromboli, 13–17 September 1999, http://www.gps.caltech.edu/classes/ge133/reading/starformation.pdf.

10 James Wolcott, "Wired up! Ready to Go!" *Vanity Fair*, 8 January 2013, https://www.vanityfair.com/culture/2013/02/quantified-self-hive-mind-weight-watchers.

11 Wolf, "The Data-Driven Life."

12 Robert Lee Holtz, "Mysterious Brain Circuitry Becomes Viewable," *The Wall Street Journal*, 23 April 2013, https://www.wsj.com/articles/SB10001424127887324235304578438811489274812.

13 See particularly, William Leiss, *Genetic Enhancement Part 1: The New Technology of Body and Mind*, https://dspace.library.uvic.ca/handle/1828/6856.

14 Patrick Tucker, "The Military Is Building Brain Chips to Treat PTSD," *Defense One*, 28 May 2014, https://www.defenseone.com/technology/2014/05/D1-Tucker-military-building-brain-chips-treat-ptsd/85360/.

15 For a full account of Virilio's theory of endo-colonization, see particularly, Paul Virilio, *War and Cinema: The Logistics of Perception*, translated by Patrick Camiller (London: Verso, 1989); and *Speed and Politics*, translated by Semiotext(e) and Mark Polizzotti (New York: Semiotext(e), 1986).

16 Tucker, "The Military Is Building Brain Chips to Treat PTSD."

17 Ibid.

18 Defence Advanced Research Projects Agency (DARPA), "Towards a High Resolution, Implantable Neural Interface: Neural Engineering Design Program Sets Out to Expand Neurotechnology Capabilities and Provide a Foundation for Future Treatments of Sensory Deficits," News Release, DARPA, 10 July 2017, https://www.darpa.mil/news-events/2017-07-10.

19 Ibid.

20 Juan Enriquez, quoted in Breanna Draxler, "Life as We Grow It: The Promises and Perils of Synthetic Biology," *Discover Magazine*, 11 December 2013, https://www.discovermagazine.com/technology/life-as-we-grow-it-the-promises-and-perils-of-synthetic-biology.

21   Bill Joy, "Why the Future Doesn't Need Us," *Wired Magazine*, April 2000.

22   Scripps Research, "Scientists Create First Living Organism that Transmits Added Letters in DNA 'Alphabet,'" *News & Views* 14, no. 15 (12 May 2014), https://www.scripps.edu/newsandviews/e_20140512/romesberg.html.

23   Ibid.

24   Heidegger, *The Question Concerning Technology*.

25   Opinno Writers, "The Internet of Things," *Opinno*, 8 November 2015, https://opinno.com/news/internet-things.

26   Dave Evans, "Why Connections (Not Things) Will Change the World," Cisco blog, 27 August 2013, https://news-blogs.cisco.com/digital/why-connections-not-things-will-change-the-world. For a full expression of Cisco's futurism, see Maciej Kranz, "IoT for Good: How the Internet of Things Is Transforming Our World for the Better," Cisco blog, 12 February 2018, https://blogs.cisco.com/innovation/iot-for-good-how-the-internet-of-things-is-transforming-our-world-for-the-better.

27   Anne Field, "Venture Capital Flocks to the 'Quantified Self,'" *The Network*, 3 June 2014, https://newsroom.cisco.com/feature-content?type=webcontent&articleId=1425860. Used with permission.

28   "The 25 Best Inventions of the Year 2013: The Edible Password Pill," *Time*, 13 November 2013, https://techland.time.com/2013/11/14/the-25-best-inventions-of-the-year-2013/slide/the-edible-password-pill/ (emphasis in original).

29   Quoted in Liz Gannes, "Passwords on Your Skin and in Your Stomach: Inside Google's Wild Motorola Research Projects (Video)," *All Things*, 3 June 2013, http://allthingsd.com/20130603/passwords-on-your-skin-and-in-your-stomach-inside-googles-wild-motorola-research-projects-video/.

30   Ibid.

31   Helen Thompson, "Gunshot Victims to Be Suspended between Life and Death," *New Scientist*, 26 March 2014, https://www.newscientist.com/article/mg22129623.000-gunshot-victims-to-be-suspended-between-life-and-death.html.

32   Ibid.

33   Ibid.

34   Ibid.

35   Ibid.

2. Power under Surveillance, Capitalism under Suspicion

1   For a major study of the entanglements of surveillance and capitalism, see, particularly, Shoshana Zuboff, *The Age of Surveillance Capitalism: The Fight for a Human Future at the Frontier of Power* (New York: Public Affairs Profile Books, 2019). From my perspective, the strength of *Surveillance Capitalism*

lies in its brilliantly meticulous belabouring of the obvious point, namely
that contemporary forms of capitalism and government in the digital
age require extensive, granular surveillance as a fundamental, necessary
condition of their basic operating logic – the organization of data points
into commercially and politically exploitable network topologies. The weakness of *Surveillance Capitalism* lies in the fact that it misses the really essential
point: specifically, surveillance can now be so intense and ubiquitous
precisely because we are living in a condition of panic surveillance, where
the real object of surveillance – that complex, unpredictable, random,
indeterminate being that is the human individual; that living soul in an
increasingly soulless world; that sometimes sparkling, often desiccated spirit
in a spiritless era – always escapes the reductionist logic of the trap of surveillance. And why not? There is no data profile, no stockpiling of digital footprints, or past data searches that can fully capture the incommensurability
of being human in the twenty-first century, always living in four-dimensional
space in a quantum zone where we are both virtual particles and finite
beings of flesh, bone, and blood at one and the same time. Confronted with
individuals who have already migrated to four-dimensional space by virtue
of being digital, surveillance capitalism only has on offer the black hole of
relational advertising and surveillance power, the crushing density of dead
mapping. Finite human beings as digital infinities have already moved on
to the fourth dimension, and that is the sure and certain source of all the
anxieties, apprehensions, and fears of panic surveillance today. For an
important description of possibilities of resistance within and against surveillance culture, see particularly Chris Hables Gray, "Veillance Society," in *The
Routledge Companion to Cyberpunk Culture*, edited by Anna McFarlane, Graham
P. Murphy, and Lars Schmeink (London: Routledge, 2020), 262–372. Hables
Gray's perspective on "veillance society" runs parallel to the critical artistic
interventions of Steve Mann, the Canadian artist, theorist, and cyberware
inventor whose tactical strategies of resistance have been globally influential,
working as they do at the cutting edge of political and aesthetic resistance to
our present deep immersion in ubiquitous streams of surveillance.

2 Dr. David Cook, personal communication (email), 3 June 2020.
3 Kashmir Hill, "NSA's Utah Data Center Suffers New Round of Electrical
   Problems," *Forbes.com*, 17 October 2013, https://www.forbes.com/sites
   /kashmirhill/2013/10/17/nsas-utah-data-center-suffers-new-round-of
   -electrical-problems/. For the most recent developments to date in 2020,
   see particularly Domestic Surveillance Directorate, "Utah Data Center,"
   https://nsa.gov1.info/utah-data-center/.
4 Howard Berkes, "Amid Data Controversy, NSA Builds Its Biggest Data Farm,"
   National Public Radio, 10 June 2013, https://www.npr.org/2013/06/10
   /190160772/amid-data-controversy-nsa-builds-its-biggest-data-farm.

5 Jamshid Ghazi Askar, "NSA Spy Center: Unsettling Details Emerge, but Director Denies Allegations," *Deseret News*, 21 March 2012, https://www.deseret.com/2012/3/21/20500226/nsa-spy-center-unsettling-details-emerge-but-director-denies-allegations.

6 Ibid.

7 Berkes, "Amid Data Controversy."

8 Jon-Michael Poff, "K-Pop Fans Are the 'White Lives Matter' and 'Whiteout Wednesday' Hashtags to Drown Out All the Racist Tweets," *BuzzFeed*, 3 June 2020, https://www.buzzfeed.com/jonmichaelpoff/K-pop-fans-flood-white-lives-matter-hashtag.

9 Alexandra Ma, "What Is China's 'Social Credit' Rating System?" *The Independent*, 8 May 2018, https://www.independent.co.uk/life-style/gadgets-and-tech/china-social-credit-system-punishments-rewards-explained-a8297486.html.

10 Lily Kuo and Helen Davidson, "Zoom Shuts Accounts of Activists Holding Tiananmen and Hong Kong Events," *The Guardian*, 11 June 2020, https://www.theguardian.com/technology/2020/jun/11/zoom-shuts-account-of-us-based-rights-group-after-tiananmen-anniversary-meeting.

11 Ma, "What Is China's 'Social Credit' Rating System?"

12 Ed Pilkington, "Washington Post Releases Four New Slides from NSA's Prism Presentation," *The Guardian*, 30 June 2014, https://www.theguardian.com/world/2013/jun/30/washington-post-new-slides-prism.

13 T.C. Sottek, "New PRISM Slides: More Than 100,000 'Active Surveillance Targets,' Explicit Mention of Real-Time Monitoring," *The Verge*, 29 June 2013, https://www.theverge.com/2013/6/29/4478572/prism-slides-surveillance-targets-real-time-monitoring.

14 Bill Chappell, "New Electronic Sensors Stick to Skin as Temporary Tattoos," *National Public Radio*, 11 August 2011, https://www.npr.org/sections/thetwo-way/2011/08/11/139554014/new-electronic-sensors-stick-to-skin-as-temporary-tattoos.

15 Ibid.

16 Ibid.

17 Ibid.

18 Ibid.

19 John Brockman, "The Technium: A Conversation with Kevin Kelly," *Edge*, 2 March 2014, https://www.edge.org/conversation/the-technium. See also Alyson Shontell, "The Next Twenty Years Are Going to Make the Last Twenty Years Look Like We Accomplished Nothing in Tech," *Business Insider*, 16 June 2014, https://www.businessinsider.com/the-future-of-technology-will-pale-the-previous-20-years-2014-6.

20 Chalmers Johnson, *The Sorrows of Empire: Militarism, Secrecy, and the End of the Republic* (New York: Metropolitan Books, 2004), 310.

21 Ibid., 278–9.

22 See particularly "Transcript of President Obama's Commencement Address at West Point," *The New York Times*, 28 May 2014, https://www.nytimes.com /2014/05/29/us/politics/transcript-of-president-obamas-commencement -address-at-west-point.html; and News Wire, "Edward Snowden on US Politics, Privacy and Technology's Threat to Democracy," *21st Century Wire*, 10 May 2021, https://21stcenturywire.com/2021/05/10/edward-snowden -on-us-politics-privacy-and-threat-to-democracy/. The article on Snowden also includes a video of the exclusive interview with Snowden conducted by MSNBC's Brian Williams and aired on 17 September 2019.

23 Alexis de Tocqueville was a French sociologist and political theorist who trav-elled to the United States in the 1830s to study American political culture. Upon his return to France, he co-authored with Gustave de Beaumont *Du système pénitentiare aux Etats-Unis, et de son application en France* (1833). A complete literal English translation has recently been published: Gustave de Beaumont and Alexis de Tocqueville, *On the Penitentiary System in the United States and Its Application to France*, translated by Emily Katherine Ferkaluk (London: Palgrave Macmillan, 2018). The authors' reflections focused on communication and political despotism, taking prison life in the United States as a key talisman of the American political system. Alexis de Tocqueville's most famous book remains, of course, *Democracy in America*, translated by Arthur Goldhammer (New York: The Library of America, 2004).

24 For a theorization of the virtual class – its genealogy, alliances, ideology, and practices – see Arthur Kroker and Michael A. Weinstein, *Data Trash: The Theory of the Virtual Class* (New York: St. Martin's Press, 1993).

25 Adam D.I. Kramer, J.E. Guillory, and J.T. Hancock, "Experimental Evidence of Massive-Scale Emotional Contagion through Social Networks," *Proceedings of the National Academy of Sciences* 111, no. 24 (17 June 2013), 8788–90, https:// www.pnas.org/content/111/24/8788.

26 Ibid.

27 Nafeez Ahmed, "Pentagon Preparing for Mass Civil Breakdown," *The Guardian*, 12 June 2014, http://www.theguardian.com/environment/earth -insight/2014/jun/12/pentagon-mass-civil-breakdown. As Ahmed reports, among the projects awarded for the period 2014–17 is a Cornell University– led study managed by the US Air Force Office of Scientific Research that aims to develop an empirical model of the "dynamics of social movement mobilization and contagions." The project will determine "'the critical mass (tipping point)' of social contagions by studying their 'digital traces' in the cases of 'the 2011 Egyptian revolution, the 2011 Russia Duma elections, the 2013 Nigerian fuel subsidy crisis and the 2013 Gazi park protests in Turkey.'"

28 Ibid.

29 Communications Security Establishment Canada, "IP Profiling Analytics & Mission Impacts," Top secret report by Tradecraft Developer, CSEC–Network

Analysis Centre, 10 May 2012, https://www.cbc.ca/news2/pdf/airports
_redacted.pdf.

30 Ibid.
31 Ibid.
32 Ibid.
33 Ibid.

3. Dreaming with Drones

1 Joseph Trevithick, "Another Syrian Terrorist Seemingly Killed by Hellfire
   Missile with Pop-Out Sword Blades," *The Drive*, 3 December 2019, https://
   www.thedrive.com/the-war-zone/31297/another-syrian-terrorist-seemingly
   -killed-by-hellfire-missile-with-pop-out-sword-blades.
2 Jonathan Allen, "Drone Malfunctions, Hits U.S. Navy Ship during Training,"
   Chicago Tribune, 17 November 2013, https://www.chicagotribune.com/news
   /ct-xpm-2013-11-17-sns-rt-us-usa-navy-drone-20131117-story.html.
3 USNI News Editor, "Navy: Six Months of Repairs to Drone-Struck Ship Will
   Cost $30 million," *USNI News*, 30 December 2013, https://news.usni.org
   /2013/12/30/navy-six-months-repairs-drone-struck-ship-will-cost-30-million.
4 See John Bigelow, Jr. *The Battle of Chancellorsville: A Strategic and Tactical Study*
   (New Haven, CT: Yale University Press, 1910).
5 Dr. Chris Hables Gray, personal communication, 5 October 2020.
6 Thomas L. Friedman, "Parallel Parking in the Arctic Circle," *New York Times*,
   30 March 2014, 11, https://www.nytimes.com/2014/03/30/opinion
   /sunday/friedman-parallel-parking-in-the-arctic-circle.html.
7 "DARPA Goes Deep: New Hydra Project to See Underwater Drones Deploying Drones," *RT*, 10 September 2013, https://www.rt.com/usa/darpa
   -underwater-drones-fleet-489/.
8 Martin Heidegger, *The Question Concerning Technology and Other Essays*
   (New Haven, CT: Yale University Press, 1977), 17.
9 "DARPA Goes Deep."
10 Mike Greenberg, "The Hydra: The Multi-Headed Serpent," *Mythology Source*,
   4 June 2020, https://mythologysource.com/hydra-serpent-greek-myth/.
11 Ibid.
12 Joan Lowy, "Drones: FAA Warns Public Not to Shoot at Unmanned Aircraft,"
   Christian Science Monitor, 21 July 2013, https://www.csmonitor.com/USA
   /Latest-News-Wires/2013/0721/Drones-FAA-warns-public-not-to-shoot-at
   -unmanned-aircraft.
13 Amnesty International, *Will I Be Next? US Drone Strikes in Pakistan* (London:
   Amnesty International, 2013), https://www.amnestyusa.org/files
   /asa330132013en.pdf.
14 For example, see the Official White House photo by Pete Souza of
   President Barack Obama and Vice President Joe Biden, along with

members of the national security team, receiving an update of the mission against Osama bin Laden in May 2011, https://www.flickr.com/photos /obamawhitehouse/5680724572/in/photostrea.

15 For Hannah Arendt's full elaboration of willing, see particularly Hannah Arendt, *The Life of the Mind* (New York, Harcourt Brace Jovanovich, 1978).

16 "A Giant Art Installation Targets Predator Drone Operators," #NotABugSplat, https://notabugsplat.com.

17 "Pakistanis Target Drones with Giant Posters of Child Victims," *Business Standard*, 8 April 2014, https://www.business-standard.com/article/pti-stories /pakistanis-target-drones-with-giant-posters-of-child-victims-114040800141 _1.html.

18 Singh, Gayeti, "I No Longer Love Blue Skies, I Now Prefer Grey Skies," *The Citizen*, 15 February 2015, https://www.thecitizen.in/index.php/en /NewsDetail/index/1/2562/I-no-Longer-Love-Blue-Skies-I-Now-Prefer -Grey-Skies.

19 Quoted in James Miller, "Vincent van Drone: They're Not Just Killing Machines Anymore." *GlobalPost*, 13 August 2013, https://www.pri.org/stories /2013-08-13/vincent-van-drone-theyre-not-just-killing-machines-anymore.

20 Ibid.

21 Quoted in Susanna Davies-Crook, "Art in the Drone Age: Remote-Controlled Vehicles Now Spy and Kill and Film Porn in Secret. So What Are Artists Doing about It?" *Dazed*, 29 November 2014, https://www.dazeddigital .com/artsandculture/article/16183/1/art-in-the-drone-age.

22 Ibid.

23 James Bridle, *Drone Shadow Handbook*, 2013, https://jamesbridle.com /works/drone-shadow-handbook. On his site, Bridle also offers "DIY Drone Shadows," a free electronic download of the *Drone Shadow Handbook* with instructions for creating drone shadows: "For some time, I've wanted to open up the project, so that anyone can draw one. With this in mind, I've created a handbook, which gives guidance on how to draw a drone shadow, including advice on measuring and materials, and schematics for four of the most common types of drone: Predator, Reaper, Global Hawk, and Hermes/Watchkeeper."

4. Robots Trekking across the Uncanny Valley

1 Associated Press, "Raw: Obama Plays Soccer with Japanese Robot," video, *Youtube.com*, 24 April 2014, https://www.youtube.com/watch?v=ag2vk6coBpI.

2 Ian Urbina, "I Flirt and Tweet: Follow Me at #Socialbot," *The New York Times*, 11 August 2013, https://www.nytimes.com/2013/08/11/sunday-review /i-flirt-and-tweet-follow-me-at-socialbot.html.

3 Ibid.

4 Ibid.

5　For B.F. Skinner's most comprehensive accounts of behaviorism, see his books *Behavior of Organisms* (New York, Appleton-Century-Crofts, 1938); and *Science and Human Behavior* (New York: Macmillan, 1953).

6　B.F. Skinner, "Operant Behavior," *American Psychologist* 18, no. 8 (1963): 503–15. https://doi.org/10.1037/h0045185.

7　B.F. Skinner, *Walden Two* (New York: Macmillan, 1948).

8　B.F. Skinner, *Beyond Freedom and Dignity* (New York: Knopf, 1971).

9　Also known as the "operant conditioning chamber," the Skinner box was developed while he was a graduate student at Harvard University.

10　"Robotic Prison Wardens to Patrol South Korean Prison," *BBC News*, 25 November 2011, https://www.bbc.com/news/technology-15893772.

11　Isaac Asimov, *I, Robot* (New York: Gnome Press, 1950).

12　For a look at some of the features of the original plan for Robot Land, see Aaron Saenz, "Welcome to Robot Land Theme Park," *Singularity Hub*, 21 February 2010, https://singularityhub.com/2010/02/21/welcome-to-robot-land-theme-park/.

13　For a description of the present state of the Robot Land project, see "Robot Land Incheon," *ThemeParX Construction Board*, https://www.themeparx.com/robot-land-incheon/. There is also now a robot theme park, Gyeongnam Masan Robot Land, in Changwon, South Korea; see Charlotte Coates, "Christie Products Power the World's First Robot Theme Park in South Korea," *blooloop*, https://blooloop.com/technology/news/christie-robot-theme-park-south-korea/.

14　"World's First Robot Theme Park to Open in South Korea," *CTV News*, 10 February 2014, https://www.ctvnews.ca/sci-tech/worl-s-first-robot-theme-park-to-open-in-south-korea-1.1679115.

15　Keane Ng, "South Korea's Giant Robot Statues to Dwarf Japan's," *The Escapist*, 8 September 2009, https://v1.escapistmagazine.com/news/view/94547-South-Koreas-Giant-Robot-Statue-to-Dwarf-Japans.

16　"Disneyland," *Visitnewportbeach.com*, https://www.visitnewportbeach.com/activities-and-attractions/disneyland/.

17　Ibid.

18　Hans Moravec, *Mind Children: Robot and Human Intelligence* (Cambridge, MA: Harvard University Press, 1988).

19　Asimov, *I, Robot*.

20　Bruce Sterling, *Crystal Express* (Sauk City, WI: Arkham House, 1989).

21　"South Korean Plans for a Robot-Themed Park Switches Back On," *New Atlas*, 24 January 2014, https://newatlas.com/south-korea-robotland-robot-theme-park/30576/.

22　"What Is the World's Robot Population?" *Robotsde.online*, https://robotsde.online/en/robots-in-the-world/.

23　Ibid.

24 William S. Pretzer, "How Products Are Made, Vol. 2: Industrial Robots," http://www.madehow.com/Volume-2/Industrial-Robot.html.

25 Oliver Wainwright, "SociBot: The 'Social Robot' That Knows How You Feel," *The Guardian*, 11 April 2014, https://www.theguardian.com/artanddesign /2014/apr/11/socibot-the-social-robot-that-knows-how-you-feel.

26 Sigmund Freud, "The Uncanny," http://web.mit.edu/allanmc/www/freud1 .pdf. See also Sigmund Freud, *The Uncanny* (London: Penguin Classics, 2003).

27 Paul Virilio, *The Vision Machine* (Bloomington: Indiana University Press, 2007).

28 Freud, *The Uncanny*, 15.

29 "The Cheetah," *Robots Voice*, 7 October 2013, http://www.robotsvoice.com /the-cheetah/.

30 Ibid.

31 Drew Guarini, "Google Acquires Boston Dynamics, Adding to Its Fleet of Robot-Makers," *HuffPost*, 14 December 2013, https://www.huffpost.com /entry/google-robots-boston-dynamics_n_4445429.

32 Eric Guizzo, "Meet Geminoid F, A Smiling Female Android," *IEEE Spectrum*, 4 April 2010, https://spectrum.ieee.org/automaton/robotics/humanoids /040310-geminoid-f-hiroshi-ishiguro-unveils-new-smiling-female-android; see also "Geminoid F," *Robots*, https://robots.ieee.org/robots/geminoidf/.

33 J. Halloy et al., "Social Integration of Robots into Groups of Cockroaches to Control Self-Organized Choices," *Science* 318, no. 5853 (2007): 1155–8, https://doi.org/10.1126/science.1144259.

34 Tom Simonite, "Robo-roach Could Betray Real Cockroaches," *New Scientist*, 9 May 2006, https://www.newscientist.com/article/dn9136-robo-roach -could-betray-real-cockroaches/.

35 "Robot 'Pied Piper' Leads Roaches," *BBC News*, 15 November 2007, http:// news.bbc.co.uk/2/hi/science/nature/7097267.stm; "Robots Manipulating Animal Behaviour," News release, European Commission CORDIS, 8 May 2006, https://cordis.europa.eu/article/id/101450-robots-manipulating -animal-behaviour.

36 J. Halloy et al., "Social Integration of Robots"; Shoshana Magnet, "Robots and Insects: Gender, Sexuality, and Engagement in 'Mixed Societies' of Cockroaches and Robots," *Women's Studies Quarterly*, 41, nos. 3/4 (2013): 38–55, https://www.jstor.org/stable/23611502. For further research, see Aisling Irwin, "Robotic Bugs Train Insects to Be Helpers, *Horizon*, 4 October 2017, https://mobile.horizon-magazine.eu/article/robotic-bugs-train -insects-be-helpers.html.

Epilogue: From the Insurgence of the Blended Mind to Gen Z

1 George Grant, *Technology and Empire: Perspectives on North America* (Toronto, ON: House of Anansi, 1969), 142.

2 David Cook, personal correspondence with the author, 27 June 2020.
3 In my estimation, the Toronto School of Communication describes the overall trajectory of a remarkable group of University of Toronto–based scholars whose writings, although dramatically different in approach, topic, and method, commonly queried the fate of culture, communication, and technology as symptomatic signs of a more generalized civilizational crisis of technological modernity. Here, profound philosophical, literary, theological, and communicative perspectives on topics ranging from media studies proper to technologies of capital accumulation, biblical codes, political economy, and classical culture were linked together by a common concern with the gathering storm of civilizational crisis. Here, Marshall McLuhan in *Understanding Media* might well brilliantly diagnose the fourfold logic embedded in technological devices, but his writing also worked to warn against the gathering storm of technologies as performing "psychic surgery" on the human sensorium. In *The Bias of Communication* and *Empire and Communications*, Harold Innis explicitly traced the genealogy of the technological maelstrom to a fatal split between time-binding and space-binding technologies of communication. It was Eric Havelock's profound insight in *Preface to Plato* that oral cultures were conservative, indexical, didactic, and preservationist in character and purpose that put paid to McLuhan's utopian aspirations for the creation of a new "Finn Cycle" of an electronic oral culture. Writing from the perspective of political economy, it was C.B. Macpherson's lasting contribution to have elaborated in text after text, from *The Real World of Democracy* to *Democratic Theory, Burke, The Life and Times of Liberal Democracy*, and *Democracy in Alberta*, the fatal implications of "possessed individualism" for a technological society premised on capital accumulation. Finally, Charles Norris Cochrane, in his classic book on Christian theology as Western metaphysics, *Christianity and Classical Culture*, descended deepest into the origins of the civilizational crisis of technological modernity when he argued that, while Christianity might have emerged as a temporary response to a deep and fatal split in Western consciousness, the weakening of religious cosmology by the technological blast has had the real effect of reawakening unresolvable contradictions in the deepest fabric of contemporary culture and society.

Index

Digital Futures

Milton Keynes UK
Ingram Content Group UK Ltd.
UKHW020038090524
442424UK00004B/300